Studien zum nachhaltigen Bauen und Wirtschaften

Reihe herausgegeben von

Thomas Glatte, Neulußheim, Germany

Martin Kreeb, Egenhausen, Germany

Unser gesellschaftliches Umfeld fordert eine immer stärkere Auseinandersetzung der Bau- und Immobilienbranche hinsichtlich der Nachhaltigkeit ihrer Wertschöpfung. Das Thema „Gebäudebezogene Kosten im Lebenszyklus" ist zudem entscheidend, um den Umgang mit wirtschaftlichen Ressourcen über den gesamten Lebenszyklus eines Gebäudes zu erkennen. Diese Schriftenreihe möchte wesentliche Erkenntnisse der angewandten Wissenschaften zu diesem komplexen Umfeld zusammenführen.

Robin Groer · Patrick J. Morrissey ·
Thomas Glatte

Historical Industrial Buildings and their Real Estate Utilisation

Comparison between Germany and the United Kingdom

Springer Vieweg

Robin Groer
Heidelberg, Germany

Patrick J. Morrissey
London, UK

Thomas Glatte
Neulußheim, Germany

ISSN 2731-3123 ISSN 2731-3131 (electronic)
Studien zum nachhaltigen Bauen und Wirtschaften
ISBN 978-3-031-89416-9 ISBN 978-3-031-89417-6 (eBook)
https://doi.org/10.1007/978-3-031-89417-6

This Springer Vieweg imprint is published by the registered company Springer Nature Switzerland AG
The registered company address is: Gewerbestrasse 11, 6330 Cham, Switzerland

Preface

As we reflect on historical industrial buildings, the majority is linked to the Industrial Revolution which had a tremendous impact on cities and places as we discover and know them today. We can still identify unique, distinctive structures and regions that were formerly utilised for industrial purposes and according to the subsequent urbanisation driven by industrialisation we may frequently find them in today's city centres, A- or B-locations, or in peripheral locations.

Because of the constantly changing needs and requirements for industrial sites, many still existing and historical industrial buildings are unsuitable for contemporary production and industrial operations.

Nevertheless, caused by a shortage of global land and restrictions on newly designated building sites, combined with trends of still ongoing urbanisation and at the same time sustainable efforts to meet international sustainability targets we must reconsider the utilisation of our existing built environments and revitalise them with new purpose. This prompts the question of how to manage historic industrial buildings, and how they can be preserved to maintain their value while simultaneously benefiting future generations.

Therefore, the publication investigates if there are differences between Germany and the United Kingdom in dealing with historical industrial buildings and potential revitalisation projects. The investigation's aim was to find out whether an international comparison may lead to strategies or approaches for future project redevelopment of historical industrial buildings.

To investigate the multifaceted issue with various subthemes, fundamental disciplines such as the Industrial Revolution with its respective national history, urban development during industrialisation, and distinctive architectural characteristics were examined and compared. A comparison with the individual national building law and heritage protection systems takes place, followed by an attempt to encircle the number of historical industrial buildings. Following upon this, there is a comparison of trends and risks for historical industrial buildings, as well as a conceptual boundary between the use of terms like "Corporate Real Estate Management" in the Anglo-Saxon and German market and context. Given those foundations, an international comparison of building law and heritage protection systems, as well as a comparison of approaches to the revitalisation of impacted buildings with the respective concepts, takes place. In addition to the unique economic and environmental circumstances encountered in historical industrial area revitalisation projects, the emphasis has furthermore been placed on social aspects and the surrounding environment.

The publication, which was generated at the Hochschule Fresenius University of Applied Sciences in Heidelberg, aims towards contributing to an expanded knowledge of the revitalisation of historical industrial buildings through an international comparison between the two nations Germany and the United Kingdom, in order to more thoroughly comprehend the obstacles identified and derive practical solutions based on the comparison.

The publication is based on an empirical investigation in which several specialists from both countries with a broad understanding of the subject participated. Our sincere gratitude goes to the experts for their valuable contributions and dedication.

Competing Interests The authors have no competing interests to declare that are relevant to the content of this manuscript.

Robin Groer
Patrick J. Morrissey
Thomas Glatte

Contents

lost places	loss of functionality of an architectonical structure and the appropriation of that structure, which can be both summarized as the afterlife of architectonical structures
NBR	National Building Record
NHLE	National Heritage List of England
NISMR	The Northern Ireland Sites and Monuments Record
NMRW	The National Monuments Record of Wales
NPPF	National Planning Policy Framework
NRHE	The National Record of the Historic Environment
RCAHMS	The Royal Commission on the Ancient and Historical Monuments of Scotland
TEV	Total Economic Value
UK	United Kingdom
UN	United Nations
UNESCO	United Nations Educational, Scientific and Cultural Organization
VAT	Value Added Tax
WWII	World War II

List of Figures

List of Tables

Introduction

1

1.1 Problem Statement

Considering our existing cities today, giving a glance into our future cities, and asking ourselves where we will live, work, and spend our leisure time within cities, the statement will be in places that already exist in the same or a similar pattern today. Although new developments are taking place, it is predicted that by 2030 only eight to nine percent of our construction activity will be new developments. This indicates that the redevelopment of already existing buildings and city districts will play an outstanding role in the future. If we look at the current existing areas and which buildings can be found, three percent of the entire property stock in Germany consists of heritage buildings according to data from the Federal Foundation for Building Culture in 2017 (Rukschcio, et al., 2019, p. 12). Considering the number of heritage buildings in the *UK*, the estimated proportion of 20% of the total stock of buildings is remarkable (Wise, Moncaster, & Jones, 2021, p. 495).

Classification and recording as a listed building in both, the sixteen federal states of Germany or, for instance, on the National Heritage List of England (NHLE) in the *UK,* reflects the historical context and value associated in both countries. The nomination and listing also symbolise the substance of preservation and the precise treatment that the individual building, ensemble, or area receives (Harzenetter, et al., 2016, pp. 12, 18; UK Public General Acts, 1990, pp. 2–3, 13).

Massive changes in the need for industrial properties have occurred as a result of the continual expansion of society, cities, and industry as influenced by causes such as digitization, *Industry 4.0,* and a variety of other influencing variables. These include structural and technical requirements for the premises

R. Groer et al., *Historical Industrial Buildings and their Real Estate Utilisation*,
Studien zum nachhaltigen Bauen und Wirtschaften,
https://doi.org/10.1007/978-3-031-89417-6_1

and accompanying infrastructure (Pfnür, Eberhardt, & Herr, Transformation der Immobilienwirtschaft—Geschäftsmodelle, Strukturen, Prozesse und Produkte im Wandel, 2022, pp. 18–9). Because new requirements for continued industrial use could not be implemented to the formerly industrial buildings or sites, or with insufficient effort, the industrial buildings were partially abounded and, in some cases, the sites and buildings were left to their fate. Weather related variables influencing the properties, as well as human activities such as vandalism, have caused some of the once industrially used industrial structures and sites to become known as 'lost places' (Bauer & Dolgan, 2020, pp. 103–06).

Merging the previously stated facts results in a new requirement for cities, their transformation, their resilience, and urban development. Due to the need for agility and adaptability within cities, new concepts and models must be developed and alternatives to previous uses of already built-up land must be derived (Pfnür, Eberhardt, & Herr, Transformation der Immobilienwirtschaft—Geschäftsmodelle, Strukturen, Prozesse und Produkte im Wandel, 2022, pp. 18–20). Although the defined worldwide proportion of built-up land is just slightly greater than one percent, more people live in urbanised regions than non-urbanised ones, and urbanisation is expected to continue to expand (Ritchie, Samborska, & Roser, 2024). As the designation of new building land is restricted, especially in cities and suburban areas, the challenge arises in innovative thought processes and the utilisation of existing structures within urban environments rather than in entirely new developments (Baum, et al., 2021, pp. 72–8). However, rethinking includes both new buildings and small-, and large-scale refurbishment. This concept is synonymous with the terms revitalisation and refurbishment. These terms might drive our future since studies have shown that Europe's building stocks nowadays already represent 80% of our buildings that will be in use by the year 2050, already exist (Gravagnuolo, et al., 2021, p. 1). Facing the global challenges of climate warming and the *UN*-goals, the net global zero targets must be achieved by 2050 (UNEP Emissions Gap Report 2023, 2023). Mitigation of global carbon dioxide emissions within the real estate industry, which is responsible for 40% of global emissions, needs to be reduced (Carlin, 2022). Thereby the usage of existing structures can contribute to sustainability by revitalising existing buildings rather than for example losing grey energy through the demolition of those (Schmitz-Morkramer, 2022, p. 469; Gravagnuolo, et al. 2021, p. 3). As we reflect on the Industrial Revolution, which had a significant impact on our cityscape, we can still identify unique, distinctive structures and areas that have previously been utilised for industrial purposes. The early development of industrial structures during the era of industrialisation, which was followed by urbanisation and numerous other changes, is the reason why those historical industrial buildings are frequently found in today's city centres, A- or B-location, or peripheral locations (Bundesstiftung Baukultur, 2019, pp. 12–6; Kirkwood, 2005, pp. 3–4).

Both Germany and the United Kingdom have economically vital cities and peripheral locations that have historic industrial structures or districts. These buildings or areas are often protected due to historical significance and thus maintained. The classification and registration as a heritage building in the sixteen federal states of Germany or on the *NHLE* reflect the value of historical context in both countries (Harzenetter, et al., 2016, pp. 12,18,20; UK Public General Acts, 1990, pp. 2–3, 13).

This prompts the question of how to manage historic industrial buildings, and how they can be preserved to maintain their value while simultaneously benefiting future generations.

1.2 Objective

Considering there are unused and abandoned historical industrial buildings in cities, the urban realm, and surrounding regions that could be developed, this study focuses on the potential and redevelopment opportunities of historical industrial buildings. While on the one hand, these historical industrial buildings reflect culturally significant buildings and sites deserving of appreciation and hence preservation, on the other hand, they are frequently situated in promising and fascinating locations that could represent significant possibilities for urban development and neighbourhood planning. Therefore, there are potential advantages arising from refurbishing the historical industrial buildings or the surrounding area. This includes re-purposing historic buildings to serve a new purpose that promotes sustainability, and social interaction, provides new housing, and enables new jobs to be developed. If meaningful utilisation concepts can be developed, these could be approaches to prevent further deterioration and avoids the need for demolition of buildings with a historical background. A comparison between the United Kingdom and Germany is of interest from a research perspective, as the *Industrial Revolution* originated in the *UK* and Germany was almost a century behind. Since the *UK*, as the initiator, has various industrial buildings and sites and also has a significantly larger proportion of listed heritage buildings, it is interesting to compare how revitalisation is handled in the *UK* in distinction to Germany.

1.3 Systematic Approach

The study investigated the research question of which distinctions exist between Germany and the United Kingdom in their approaches to handling historical industrial buildings and potential developments. A qualitative study was conducted to determine the response to the research question using expert interviews, which were carried out online via video conference. Thereby the interviews were executed online as semi-structured interviews due to physical distance. The objective is to identify possible differences and to derive recommendations for action for the respective countries that could result from different approaches. The study is outlined starting with the fundamentals in Chapter 2. This involves examining the legacy of the existing historical industrial buildings, deriving the various terms used, and establishing a common understanding for the paper, the structure of urban planning and heritage protection, special characters of historic industrial buildings, and trends affecting the revitalisation of the properties. This is followed by a presentation of the qualitative study carried out with the analysis of the interviews conducted. An attempt is afterwards made to derive approaches from the differences between the two countries based on the comparison of handling and redeveloping historical industrial buildings.

Explanation and Fundamental Principles 2

2.1 Industrial Revolution

Prior to discussing the redevelopment of historical industrial buildings, it is essential to consider the background of their heritage. The *Industrial Revolution* is one of the major transformative periods of global history. Identifying the exact commencement of the *Industrial Revolution* poses challenges since it was a series of events rather than a single incisive event. It was a long-lasting and gradually implemented process of agricultural and industrial processing, which consequently led to numerous significant and lasting economic, social, architectural, and urban changes predominantly by implementing new technologies and inventions over decades (Cartwright, 2023; Stearns, 2021, pp. 42–49). This paper explores the timeline of the *Industrial Revolution*, referencing various authors' statements from the mid-eighteenth to the mid-nineteenth centuries (Morgan, 2011, p. 4; Hudson, 1989, p. 6; Allen, 2009, pp. 2–3; Hahn, 2011, p. 1).

Fundamentally, Britain is credited with initiating the *Industrial Revolution* during the 1760s and then spreading industrialisation toward America and Continental Europe (Allen, 2009, pp. 135–136; Stearns, 2021, pp. 42–47). The transformation of agrarian institutions, with the enclosure of open fields, and the establishment of large-scale capitalist farms (Allen, 2009, pp. 57–59) was the fundamental requirement that resulted in the expansion of the urban population, increased production management, and economic development (Jackson, 1985, pp. 349–50). Britain's labour market was characterised by a significant amount of manual labour and extraordinarily high wages during the eighteenth century (Morgan, 2011, p. 3). At the same time, considerably low energy costs caused by abundant coal resources and a large-scale coal industry, which became crucial

R. Groer et al., *Historical Industrial Buildings and their Real Estate Utilisation*, Studien zum nachhaltigen Bauen und Wirtschaften,
https://doi.org/10.1007/978-3-031-89417-6_2

for mechanisation and industrial developments, were predominant (Allen, 2009, pp. 80–83; Morgan, 2011, pp. 3–4).

Aligning these two circumstances labour started being mechanised over the next decades by creating labour-saving machinery which balanced high wages with cheap energy (Allen, 2009, p. 136). Since the beginning of the *Agricultural Revolution*, between 1600 and 1760, the city of London, together with the cotton industry, played a significant role in the *Industrial Revolution*. London's influence was founded as a financial market on overseas trade, which Britain established early on with Eastern countries (Clarkson, 1990, p. 4). Compared to other countries, Britain's advantage was based on their harbours and canal system with narrowboats which had been fundamentally for national transportation of raw materials imported through London and transported to the north, up to Manchester (Hudson, 1989, pp. 66, 70, 109).

After the insight into Britain's pre-industrial era, the main inventions must be considered with their impact on the social structure, properties, and culture which together symbolise the significant changes of this time (Allen, 2009, pp. 6–7). The inventions can be structured in macro- and micro-inventions. The steam engine by Thomas Newcomen invented in 1712, was the first micro-invention of its kind (Allen, 2009, pp. 140, 156). More significant inventions were J. Hargreave's *Spinning Jenny* in 1764, R. Arkwright's *Waterframe* in 1769, and S. Crompton's *Spinning Mule* in 1770 which transformed the cotton industry, becoming the icon of British industrialisation. The inventions transitioned textile production from domestic handcrafting towards mechanised factory mills while increasing productivity and reducing costs. This shift enabled Britain to later dominate the cotton and cloth industry (Allen, 2009, pp. 183–84; Stearns, 2021, p. 32). Newcomen's steam engine was improved by James Watt in 1769, making it more efficient and adjusted for industrial purposes such as powering machines. Establishing a new manufacturing process improved output and led to the first factories being built (Allen, 2009, pp. 166–72). Abraham Darby's coke blast furnace in 1709 (Allen, 2009, pp. 217–18) and Henry Bessemer's steel-making technique later in 1856 (Stearns, 2021, pp. 39–41) empowered the production of iron and steel.

These materials were crucial for the construction of railways, steamships, and machinery, which additionally supported industrial expansion (Hudson, 1989, p. 211). With the available resources, not only the machinery but also the construction industry was transformed at the end of the eighteenth century by constructing the first multi-story cotton mill. This launched the concept of iron-framed construction with a far-reaching impact on the architectural and building industries. *Shrewsbury Flaxmill Maltings* was the first construction to implement this method. Furthermore, new technologies enabled constructions to be illuminated by gas a few years later (Clarkson, 1990, p. 50). An additional macro-invention that led to an entirely novel phase of transportation, was the first functioning railway between Liverpool and Manchester, which was established in 1830 (Stearns, 2021, p. 40).

Shifting the perspective from the British perspective to the globalisation of the *Industrial Revolution*, Germany experienced a delay in entering the *Industrial Revolution* era. This was caused by the *Napoleonic Wars* and the prohibition of economic expansion through continental trade restrictions (Ziegler et al., 2012). The *Industrial Revolution* in Germany was comparable to the *UK*'s, beginning with the essential *Agrarian Revolution* in the late eighteenth century. In Germany, it was characterised by the shift in estates characterised by manor, the breaking of the medieval labour guilds, and the dissolution of serfdom which had been essential for the economy to develop (Ziegler et al., 2012, pp. 18–20, 27–8). These developments led to domestic trade being able to establish in the German market, territorial land consolidation, and the standardisation of the German economic area being able to operate on the international trade market (Hahn, 2011, p. 10).

After the pre-industrial period, Germany imported the first machinery from the *UK* and started to build factories between 1850 and 1870 (Allen, 2009, p. 2). Germany began to specialise in the field of heavy industry rather than in the woollen and spinning industry. Key factors were the development of iron production and subsequently the implementation of mechanical engineering and the railway network across the country (Ziegler et al., 2012, p. 51). The high demand for energy was subsequently met by hard coal production. Both hard coal and iron have been important industries, and their production sites and legacies are still evident today, for instance, the Ruhr (Hahn, 2011, p. 27).

In Germany, the end of the nineteenth and beginning of the twentieth century often has been stated as the second *Industrial Revolution*, when crafting and tinkering ended, followed by the implementation of new technologies associated with mass consumption. This resulted in standardising and the first automation of manufacturing processes while lowering material input and raising output with improved standards of quality of goods (Ziegler et al., 2012, pp. 101–103). Further accelerators for the electrification of manufacturing processes and Germany's rise as a leader in electrical and chemical industries included the adoption of technologies invented in America, such as telephone networks and electrical lighting (Ziegler et al., 2012, p. 102; Josephson & Conot, 2024). Apart from these two industrial sectors, the mechanical engineering of locomotives started expanding, followed by the automobile sector under the influence of Ford's initial assembly-line innovation (Ziegler et al., 2012, pp. 128–129; Binder & Rae,

2024). According to these developments, the third phase of the *Industrial Revolution* was associated with the digital revolution and digitalisation began (Voigt, 2018). The *Industrial Revolution* had long-lasting effects, including the construction of numerous factories and industrial sites (Rukschcio et al., 2019, pp. 14–15). In addition to the factories already located in the city centres, those factories situated in suburban areas became integrated into the main urban area following urban sprawl (Ziegler et al., 2012, pp. 38–39). However, many of these sites were repurposed during the World Wars, often serving as production facilities for military equipment. Furthermore, post-war destruction and subsequent economic changes led to the massive decline of many industrial sites being bombed or demolished during or after the war (Blomvall, no date).

2.2 Conceptual Delimitations of Terms

Focusing on industrial buildings, it is important to clarify the specific interpretation of this phrase, as there is no common standard.

Approaching the term from the *UK* perspective, two expressions are generally used to describe properties utilised by companies. Firstly, *commercial real estate* is a wide-ranging term for all properties that are used for a business-related purpose and furthermore for buildings that have at least four residential units. It includes the categories of office, retail, industrial real estate, and apartment buildings. Accordingly, the definition of this term can be seen as an asset class with income-generating purposes (De Roos, 2008, pp. 1–3). Secondly, *industrial real estate* can be considered for all categories of manufacturing facilities, logistics, distribution centres, and transformational properties (Statista, no date). It is associated with properties that provide a workplace of any kind and can be distinguished from residential buildings. Accordingly, *industrial real estate* can be classified as non-residential real estate with a utilisation exclusively for business purposes (Cambridge Dictionary, 2024).

Approaching the definition of *commercial* and *industrial real estat*e from a German perspective, there is a more detailed definition of various types and categories of properties. Initially, a distinction between undeveloped and developed properties is undertaken. Subsequently, within developed sites, a classification of *commercial properties* can be found, which was expressed in line with the Anglo-Saxon market, and has replaced the previously existing term *corporate property* in order to homogenise it. As depicted in *Fig. 2.1*, the term *corporate property* as a subdivision can be found (Ache et al., 2016; bulwiengesa AG, 2014; adapted by Glatte, 2022). Regarding the German valuation perspective, the classification of *industrial property* is allocated to the *special property* in a broader sense. Thus, the definition includes logistic properties, factories, industry, refineries, chemical plants, steel plants, and business parks (Bienert & Wagner, 2018, p. 9). Differentiating *commercial* and *industrial real estate,* the descriptions within land use plans are helpful. The legal text of §8 *BauNVO* responsible for designating *commercial real estate zones* signifies that commercial zones primarily serve for the establishment of businesses that do not cause significant nuisance (BauNVO, 2022, p. 341). The following legal text, §9 *BauNVO* responsible for *industrial real estate* implies that the *industrial real estate zones* serve for the settlement of enterprises which do not meet the requirements of other designated zones (BauNVO, 2022, p. 342). Accordingly, *industrial real estate* includes buildings and industrial plants which might cause significant nuisance due to pollution, noise et cetera. The definition of *industrial real estate* as an asset class indicates a broad range of building occupants from several sectors with specific manufacturing needs. Thus, industrial sites are often not only used for a single purpose but combine a broad range of typologies within one location including production buildings, logistic buildings, offices, transformational real estate, and business parks (bulwiengesa AG, 2014).

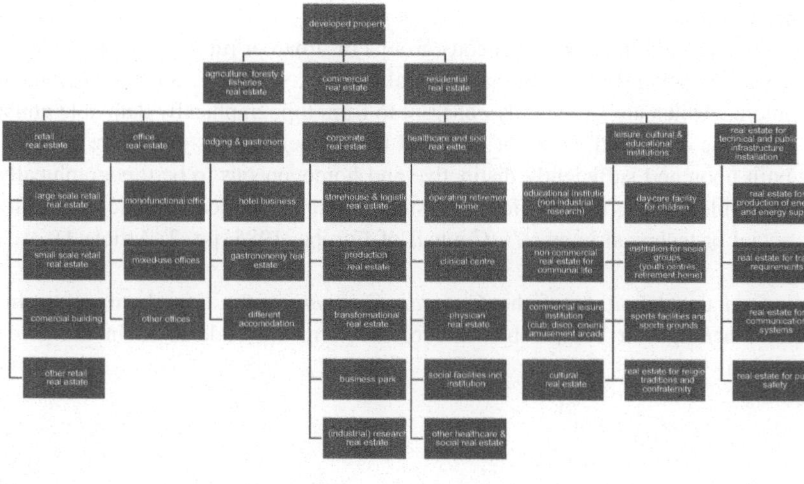

Fig. 2.1 Asset class developed properties, own illustration according to (Ache et al., 2016, pp. 16–25) converted with (bulwiengesa AG, 2014, p. 5) by Prof. Dr. Ing. Thomas Glatte

To ensure a consistent understanding throughout the paper, the terms *commercial real estate* and *industrial real estate* are employed by the German definition. Furthermore, the use of the term *industrial buildings, construction,* or any other use comparable use is based on the German definition.

Furthermore, a clarification of the terms *heritage* and *architectural heritage* is necessary due to different meanings and understanding of the terms.

Since both countries, Germany and the United Kingdom, ratified the *Granada Convention* in 1985, which was declared the *Convention for the Protection of the Architectural Heritage of Europe* they have pursued related goals. The *Granada Convention* thereby was signed by thirty-two participating *EU* states to establish a concept of active heritage conservation at an international level and underline the significance of a shared policy for safeguarding Europe's architectural legacy and increasing public awareness of its relevance (Pickard, 2002, pp. 349–350).

In order to have consistent terminology for the conservation laws, the term *architectural heritage* was defined in Article One. Accordingly, "… the expression 'architectural heritage' shall be considered to comprise the following permanent properties:

1. monuments: all buildings and structures of conspicuous historical, archaeological, artistic, scientific, social or technical interest, including their fixtures and fittings;
2. groups of buildings: homogeneous groups of urban or rural buildings conspicuous for their historical, archaeological, artistic, scientific, social or technical interest which are sufficiently coherent to form topographically definable units;
3. sites: the combined works of man and nature, being areas which are partially built upon and sufficiently distinctive and homogeneous to be topographically definable and are of conspicuous historical, archaeological, artistic, scientific, social or technical interest." (Council of Europe, 1985, pp. 2, Article 1).

Based on the definition of the *Granada Convention,* the term heritage, architectural heritage, or comparable uses of the term are understood in the given definition.

2.3 Heritage Protection

2.3.1 UNESCO World Heritage

Conventionally, national governments and their public authorities are responsible for heritage protection prioritising the safeguarding of historically significant buildings. Throughout the past decades, heritage protection has seen a broadening of its scope, owing to factors such as regionalisation, globalisation, and collaborative efforts (Zwegers, 2022, pp. 3–4). The highest authority for the protection of historically significant artefacts is *UNESCO*. In 1942, European governments met in the *UK* for the *Conference of Allied Ministers of Education*, seeking ways to rebuild education systems while *WWII* was still proceeding. Immediately after the war ended, the conference gathered to establish an organisation dedicated to promoting peace and establishing the intellectual and ethical solidarity of mankind to avert another world war. The *UNESCO* was thus created by the *UN* in 1945, post *WWII* times to foster international cooperation with the main goal in education, sciences, culture, communication, and information. Its primary goal is to promote knowledge sharing and the free flow of ideas to enhance mutual understanding (UNESCO Constitution, 2023). *UNESCO* has an organisation, the *International Council on Monuments and Sites (ICOMOS)*, acting as their advisor to the *World Heritage Committee* to establish whether newly nominated sites should be added to the *UNESCO World Heritage List*. The list has existed since 1978 and contains heritage sites of remarkable international value that must be protected. Besides that, the *ICOMOS* is in charge of monitoring already enlisted sites, assessing their condition, and evaluating preservation methods (Zwegers, 2022, pp. 8–9).

Paragraph 49 of the policy compendium by *UNESCO* can be accessed to gain a better understanding of the concept of world heritage: "Outstanding Universal Value means cultural and/or natural significance which is so exceptional as to transcend national boundaries and to be of common importance for present and future generations of all humanity. As such, the permanent protection of this heritage is of the highest importance to the international community as a whole." (UNESCO World Heritage Centre, 2019, pp. 20, No. 49.). The *World Heritage List* contains monuments, groups of buildings, and sites. Monuments include paintings, memorable sculptures, and architectural masterpieces. Groups of buildings are collections of detached or linked structures due to their architectural design, their uniformity, or their location within scenery. Sites are characterised as the creation of individuals or the combined works of humans and the environment, and areas enclosing archaeological evidence (UNESCO World Heritage Centre, 2019, p. 19). If a community, city, or association considers a site, group of buildings, or a monument to fulfill the requirements according to the definition, they must follow a strict application procedure and format, as outlined in the German abbreviated form. First of all, the national specialised authorities must be approached. The significance must be verified by the first authorities before the application is passed on to the national authority and might then be added to a suggested site list of the country. After the official application documents have been prepared the nomination at the *ICOMOS* will follow and a decision will be undertaken by the *World Heritage Committee.* The inclusion on the list may be permitted or prohibited, depending on the decision of the committee (Eisenmann, 2017, pp. 22–23).

A listing on *World Heritage List* entails several duties for the nation and the authorities which are set out in protection and management guidelines with the main components:

- "Legislative, regulatory and contractual protection measures
- Boundaries for effective protection
- Buffer zones
- Management System
- Sustainable use" (UNESCO World Heritage Centre, 2019, pp. 4, 29–33; II.F).

The *UNESCO* list currently encompasses 1,119 heritage sites in 168 countries, 48 of them being transnational. Heritage sites are categorised as cultural, natural, and mixed properties with 56 of them being additionally included in the *List of Heritage in Danger*. The *World Heritage List* includes 52 sites in Germany and 33 sites in the United Kingdom of Great Britain and Northern Ireland. Thus, both countries have protected industrial heritage sites (Table 2.1) (UNESCO-World-Heritage-Centre, no date).

2.3.2 United Kingdom

2.3.2.1 Urban Planning

The *Industrial Revolution* ushered in the beginning of an era characterised by the emergence of factories and industrial sites, which strongly influenced the urban landscape and social lives. Referring to the first mentioned factories, they must be differentiated from what we associate factories nowadays. During this period, production facilities were designed for labour, but there was still a strong interdependence between housing, leisure time, and labour (Stearns, 2021, p. 32).

Table 2.1 UNESCO World Heritage List—Industrial Heritage Sites; Own compilation according to (UNESCO World Heritage Centre 1992–2024, no date; UNESCO World Heritage Centre 1992–2024, no date)

Germany	United Kingdom of Great Britain and Northern Ireland
Fagus Factory in Alfeld (2011)	Blaenavon Industrial Landscape (2000)
Mines of Rammelsberg, Historic Town of Goslar and Upper Harz Water Management System (1992, 2010)	Cornwall and West Devon Mining Landscape (2006)
	Derwent Valley Mills (2001)
Speicherstadt and Kontorhaus District with Chilehaus(2015)	New Lanark (2001)
Völklingen Ironworks (1994)	Saltaire (2001)
Zollverein Coal Mine Industrial Complex in Essen(2001)	

After a short period, massive problems arose due to industrial sites and the overcrowding of cities took place. At the same time, unplanned suburbs emerged after companies' sites were coexistent. These were often situated close to the housing of the employees due to the lack of urban development plans existing. Aside from these developments, cities experienced overloading, severe pollution, and health issues due to a lack of efficient infrastructure (Morgan, 2011, pp. 28–29, 121–122). As a result, public discourse in the mid-nineteenth century sparked debates, highlighting the evident lack of urban planning. Accordingly, the need for controlled building regulation grew essential, leading to the first instances of town planning being discussed in political debates, resulting in the initial *The Housing, Town Planning,* etc. *Act 1909* (Sutcliffe, 1988, pp. 291–294). Thus the *UK,* including the four countries England, Northern Ireland, Scotland, and Wales got the same fundamental planning system structure. However, their operations differ in terms of details and methodology. The four national planning systems all have a plan-led approach which refers to the overarching development plan. This outlines which plannings should and which plannings should not receive authorisation. It specifies the protection of land and emphasises the public interest in environmental protection. Since Wale's fundamental legislative system is similar to the English, the following statements focus on the English (Winter et al., 2016). The highest public building law requirements from *England* were set by the Government with the *National Planning Policy Framework* (NPPF) which states economic, social, and environmental criteria for development undertaken. This policy must be considered by the following tiers of municipal and local authorities. There are usually three tiers which include county council, district/borough or city council, and town councils (Fig. 2.2).

The superordinate framework must be considered in local or neighbourhood spatial planning. Basic urban development mainly concerning the local areas will be undertaken by the authorities on their responsibility. However, some far-reaching or infrastructure projects with national significance must be authorised by a higher body (Department for Communities and Local Government, 2014, pp. 4–7). Local plans are primarily used to collaborate with local communities, providing an overview of future urban planning. Neighbourhood planning is a comparatively new planning approach that allows the local population to contribute their interests and suggestions to the planning process. This approach has been in place since the *Localism Act 2011* and requires review with local strategic planning following the submission of applications. Applications must go through several review bodies and then be approved by the residents in a referendum before they can become legally effective (Department for Communities and Local Government, 2014, pp. 11–12).

2.3.2.2 Building Protection

In the *UK*, while the general principles of the system are consistent across England, Wales, Scotland, and Northern Ireland, there are different regulations specific to each country regarding the preservation of heritage (Table 2.2). Due to the fact that England and Wales got a common legislative structure (Taylor & Sorenson, The Building Conservation Directory 2014, 2014) and both countries together have almost two-thirds of the land area, and 85% of the population, more detailed explanations will refer to England, unless otherwise stated (Urmersbach, 2024).

The roots of heritage protection in England can be traced back to *The Ancient Monuments Protection Act, 1882* which was the first act by parliament concerning the protection of heritage sites (UK Parliament, no date). The act initially defined the responsibility and treatment of heritage concerned in terms of maintenance. It contained the first list of ancient monuments categorised by the four nations and included around fifty monuments that were looked after by appointed inspectors. In addition, the first penalties for violating and damaging listed monuments were specified (Act of Parliament, 1882, pp. 1–9). Subsequently, several acts like *The Ancient Monuments Consolidation* and *Amendment Act of 1913* followed and expanded the law, increasing the scope of heritage protection. Still, the focus was mainly on the heritage protection of the Middle Ages castles, stone circles, and similar structures (English Heritage, English Heritage, no date). Thus, the first

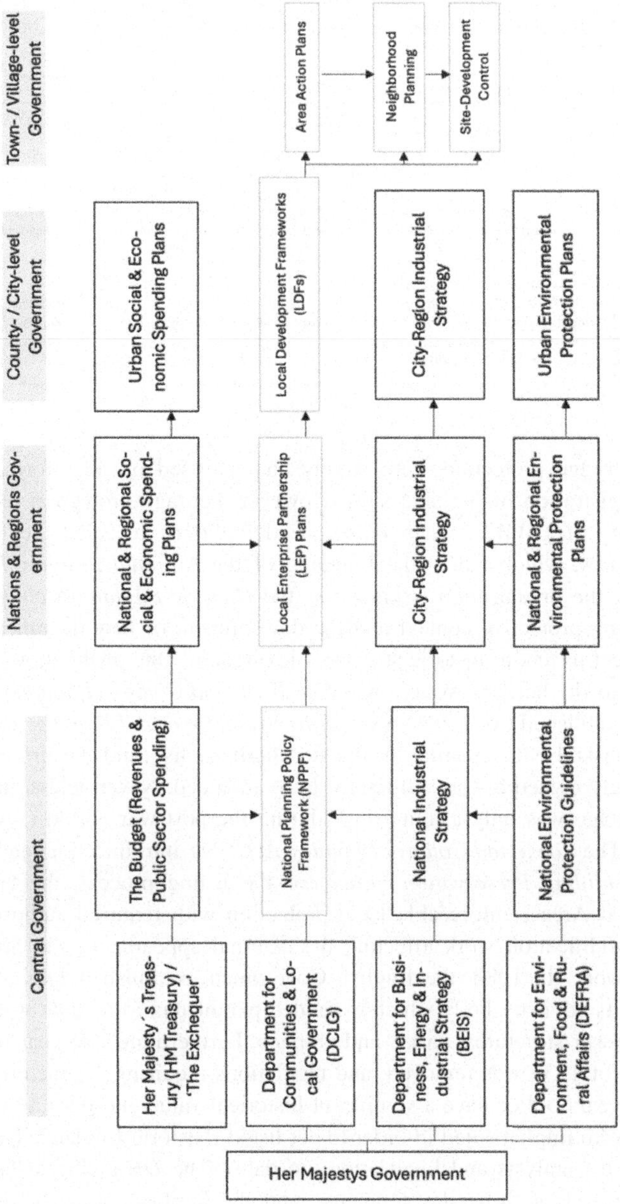

Fig. 2.2 England and Wales' hierarchical planning system, own illustration according to (Hacking & Flynn, 2018, p. 4)

Table 2.2 Overview over Heritage Protection subdivided by countries from the *UK* and legislative basis; own illustration according to (Taylor & Sorensen, The Building Conservation Directory 2021, 2021, p. 25)

	PRIMARY LEGISLATION	GOVERMENT POLICY AND GUIDANCE	STATUTORY AUTHORITY
ENGLAND	*Planning (Listed Buildings and Conservation Areas) Act 1990*	*The National Planning Policy Framework*	Historic England
WALES		*Planning Policy Wales, Technical Advice Note 24: The Historic Enviroment* and *Managing Change to Listes Buildings in Wales*	Cadw
SCOTLAND	*Planning (Listed Buildings and Conservation Areas) (Scotland) Act 1997*	*Historic Enviroment Policy for Scotland* and the guidance notes in the *Managing Change in the Historic Enviroment: Use and Adaptation of Listes Buildings*	Historic Enviroment Scotland
NORTHERN IRELAND	*The Planning Act (Northern Ireland) 2011* *The Planning (General Development Procedure) Order (Northern Ireland) 2015* *The Planning (Listed Buildings) Regulations (Northern Ireland) 2015*	*Strategic Planning Policy Statement 2015*	Historic Enviroment Division, Department for Communities

attempts at heritage protection were largely characterised by an informal collection and registration of heritage sites. However, heritage protection received major attention after *WWII*, when numerous demolitions revealed the loss of historic structures, which led to the founding of the *National Building Record* (NBR). Today, the organisation is known as the *Historic England Archive.* Due to the expanding protective context and the development of new documentation tools, the project to record historic structures is ongoing. One of the most significant changes to the heritage system was *The Town and Country Planning Act of 1944*, which established the new *Minister of Town and Country Planning.* The new ministry was tasked with creating an updated heritage list, and the first reform allowed privately owned historic structures to be included. Nevertheless, the protection of heritage was only a minor task during the post-war period (Zwegers, 2022, p. 37). The *Town and Country Planning Act 1994* introduced amendments such as the *Building Preservation Notice* for the listing process, the creation of Conservation Areas, enforceable legal protection with required authorisation for planned modification work affecting the external appearance, and fines for unauthorised work. In 1983, the English Government established *English Heritage (EH)* as its advisory body which expanded parameters to include war sites, churches, post-war structures, parks, and gardens. Furthermore, the year of construction limit of 1939 was removed, and regulations requiring a structure to be at least thirty years old or have a significant historical value and pose a threat to conservation were implemented in order to get listed (Historic England, Timeline of Conservation Catalysts and Legislation, no date). The *UK*'s current heritage policy is the result of those developments, and the heritage office became an

Table 2.3 Designation of Heritage Protection in England, own illustration according to (Taylor & Sorenson, The Building Conservation Directory 2014, 2014)

Designation	Works requiring consent	Consent required
Scheduled monument	All works including demolition, alterations and repairs	Scheduled monument consent (SMC)
Listed building	All demolition work and alterations which affect its character as a listed building - this includes works to the interior, objects and structures fixed to the building, and objects and structures within its grounds built before 1948 or, in Northern Ireland, 1973.	Listed building consent (LBC)
Unlisted building in a conservation area	Demolition only ("substanial" not partial demolition)	Conservation area consent (CAC) or, in England, planning permission
	Some external alterations to houses, which elsewhere would be permitted by right, may require consent under an Article 4 direction	Planning permission
	In addition to the above consents, development affecting the exterior of a heritage asset may also require planning permission	

active partner and promoter, based on the experience gained (Zwegers, 2022, pp. 36–37). Discussing the conservation of architectural heritage in the *UK*, there are different designations for the protection status regarding scheduled ancient monuments, listed buildings, registered parks and gardens, conservation areas, world heritage sites, locally listed buildings, wreck sites, and battlefields (Society for the Protection of Ancient Buildings, no date). Nevertheless, this study focuses on the three common categories depicted (Table 2.3).

England's earliest form of heritage protection was scheduling buildings. Tracing back to *The Ancient Monuments Protection Act 1882* only applied to ancient constructions, but its official application did not occur until 1913. Although scheduling is within the responsibility of *Historic England*, it can be initiated by everyone and applications need to be sent to the responsible *Secretary of State for Culture, Media and Sport* (DCMS). Considering the characteristics of applied buildings for the *NHLE*, there are twenty categories which have two hundred subdivisions. A building can be considered for listing if it is either found in the inclusion in one of the two hundred categories, has a relevant historical background, or is a rare example of its construction type (Historic England, Scheduled Monuments, no date). The two hundred categories range from agricultural buildings, and military structures to street furniture, garden and parks, and industrial buildings (Historic England, Listing Selection Guides, no date). Buildings constructed before 1700 with few examples existing, are regularly listed. Furthermore, most buildings constructed between 1700 and 1850 are likely to be listed, but consideration is required especially for those built after 1945. Buildings

Table 2.4 Listed Building Grades/Categories *UK*, own illustration according to (Taylor & Sorensen, Building Conservation Directory 2023 30th Edition, 2023, p. 26)

LISTED BUILDING GRADES/CATEGORIES (and proportions in each)			
England	Grade I (2.5 %)	Grade II* (5.5 %)	Grade II (92 %)
Wales	Grade I (2 %)	Grade II* (6 %)	Grade II (92 %)
Scotland	Category A (8 %)	Category B (50 %)	Category C (42 %)
Norther Ireland	Grade A (2.5 %)	Grade B+ (6.5 %)	Grade B1/B2 (91 %)

less than thirty years old are frequently not considered of special architectural or historic interest as they have not yet withstood the test of time. In England, listed buildings are categorised as *Grade I* buildings for exceptional interest, *Grade II** for buildings of more than special interest and particular importance, and *Grade II* for buildings of special interest which is the most awarded grade (Historic England, Listed Buildings, no date). Although the grade or category of the listing varies by country (Table 2.4), built heritage is uniformly protected by law (Taylor & Sorensen, The Building Conservation Directory 2021, 2021, p. 25).

Whether an applied building will be granted a listed building grade is dependent on the decision by the *DCMS* verifying the concerned age, rarity, documentation, group value, condition, fragility, diversity, and potential (Historic England, Scheduled Monuments, no date). Once architectural heritage has been assigned a grade or category, it will be included on the statutory list of buildings. The databases on which the listings are publicly available are The National Heritage List of England (NHLE), The Northern Ireland Sites & Monuments Records (NISMR), The Royal Commission on the Ancient and Historical Monuments of Scotland (RCAHMS), and The National Monuments Record of Wales (NMRW) (Taylor & Sorenson, The Building Conservation Directory 2014, 2014).

The listing on the *NHLE* encompasses the entire structure, including its interior and exterior, as well as any supplementary buildings or attached structures from years prior to 1948. Thus, all parts of the building and fixtures and fittings are protected unless explicitly excluded in the registration text for the *NHLE*. This implies that all building components, such as staircases and fireplaces are included. Items such as paintings are generally excluded unless they are part of the architectural or cultural heritage. This requires individual examination and might be a legal grey area (Morrice, Listed Building Consent, 2021, p. 13). In addition, a second listing exists that emphasises buildings that are at risk of being lost through being vacant, deterioration or inadequate development. It is based on *The Heritage at Risk* (HAR) programme by *HE* and is annually updated on the *Heritage at Risk Register* which evaluates the built heritage's conditions. *HAR* conducts status reports and is focused on lowering the risk to heritage. They consult owners with funding opportunities, help identify structural measures, and help carry out conservation (Historic England, What is the Heritage at Risk Programme?, no date). Heritage protection is subject to a special planning system in order to preserve it for future generations. Whenever an alteration of an architectural heritage building is planned, *Listed Building Consent* (LBC) is necessary as soon as planned work might affect the perception of the construction. The necessity of *LBC* is due to the *Planning (Listed Buildings and Conservation Areas) Act 1990*, which states "... no person shall execute or cause to be executed any works for the demolition of a listed building or for its alteration or extension in any manner which would affect its character as a building of special architectural or historic interest, unless the works are authorised." (Acts of Parliament, 1990).

Derived from the *Planning Act 1990*, some alterations certainly do not require authorisation, other alterations where *LBC* may be necessary, and measures where it is required. In the same series, they include gutter cleaning, new lighting based on the kind, appearance, and scope of electrical restoration, and destruction or expansion of building structures. If there is uncertainty about whether *LBC* is necessary or not, expert advice shall be obtained or a pre-application shall be sent in order to discuss the compliance of planned measures and avoid criminal offences by undertaking non-authorised alterations (Morrice, Listed Building Consent, 2021, pp. 5–6). Furthermore, the local authority may require general planning permission besides the *LBC* if, for instance, the exterior appearance of the building is affected. In contrast to the additional regulations, a *General Permitted Development Order* (GPDO) can be applied in a scaled-down form for built heritage sites (Morrice, Listed Building Consent, 2021, p. 11).

Conservation Areas determined by local planning authorities designate protected areas based on their unique architectural or historic interest (Department for Levelling Up, Housing and Communiti, 2023, p. 57 NPPF paragraph 197). *NPPF*-based conservation areas shall be designated to emphasise architectural traditions and settlement patterns of characteristic townscapes. Planning authorities within local development and neighbourhood plans determine significant potential in the vast number of 10,000 current conservation areas with at least one in every local authority (Morrice, Conservation Area Appraisal, Designation and Management, 2019, pp. 1–3, 7–8). Special protection for every building within the area results in the preservation of the external appearance. Conservation areas, for instance, include centres of historic cities, fishing and mining villages, eighteenth to twentieth-century suburbs, historic transport links such as canal sections, and industrial heritage sites (Historic England, Designating and Managing a Conservation Area, no date). They provide regulations for local authorities regarding needed consent for altering or developing properties. They include the demolition of not listed buildings, and restrictions on advertisements displayed on facades (Morrice, Conservation Area Appraisal, Designation and Management, 2019, p. 4). Furthermore, *Article 4 Direction* by *GPDO* is available to the authorities. Thus minor developments such as domestic alterations and extensions which generally would be permitted can be restricted and are subject to permission if the local authorities deem it necessary in certain areas (Morrice, Conservation Area Appraisal, Designation and Management, 2019, pp. 30–33, 40–41). The *UK* government provides tax concessions regarding capital taxation for architectural heritage which include capital gains tax, and inheritance tax which replaced estate duty (HM Revenue & Customs—GOV.UK, 2022, p. Chap. 2). Tax reliefs are only granted if certain conditions are met, such as the eligibility of the building. Thereby listed buildings are pretty likely to be eligible for heritage status but it is not guaranteed and a further investigation of *English Heritage*, the *Northern Ireland Environment Agency*, *Historic Scotland*, and *Cadw* might be necessary for confirming the status (HM Revenue & Customs—GOV.UK, 2022, p. Chap. 6.1). However, this must be examined case by case and is not the subject of this thesis as well as the regulations for the relief of inheritance tax which are subject to specific conditions such as maintenance and use of the buildings and possible sales (HM Revenue & Customs—GOV.UK, 2022, p. Chap. 3). In addition to tax reliefs granted by the government, there are public funds available. Those funds might be specialised in the conservation of architectural heritage and support maintenance work for historically relevant structures (HM Revenue & Customs—GOV.UK, 2022, p. Chap. 3.23). Besides tax reliefs, there are also institutions with a notable example supported by the government, *The National Lottery Heritage Fund,* which is an independent public organisation under the authority of *DCMS*, the primary funder of *UK Heritage*, which collects money from National Lottery participants. As a part of The Heritage Fund, founded in 1994, it has awarded £9.2 billion of National Lottery and other funding to over 52,000 projects (The National Lottery Heritage Fund, no date).

2.3.3 Germany

2.3.3.1 Urban Planning

Caused by the *Industrial Revolution*, urbanisation and massive growth of cities took place during the nineteenth century. Intensive land utilisation and a partial deterioration of social structures were the results of urbanisation. These circumstances led to the necessity for a structure of urban development. Responding to this, the Baden alignment law "*Badisches Fluchtliniengesetz*" from 1868 and the Prussian alignment law "*Preußisches Fluchtliniengesetz*" from 1875 were implemented. These developments are closely linked to the origins of contemporary urban planning systems and the building law (Glatte, Entwicklung betrieblicher Immobilien, 2014, p. 58). Focussing on the structure of land planning and utilisation in Germany there are different hierarchical levels (Fig. 2.3). The highest level of the hierarchy can be assigned to the European *EUREK* system even though it has no legally binding effect. Nevertheless, it focusses on society, economy, and environment and provides a sustainable *European Spatial Development* guideline (Ewringmann, Perner, & Schön, 2000, p. XV). On a national level, the German government provides a framework concept for all federal states as part of the public building law. Provided by the government, it is a spatial development guidance principle and determines ideologies for spatial planning. Each federal state acts independently and adapts the national spatial planning, converting it into its spatial system within the federal state (Krüper, 2016, pp. 357–358). Furthermore, federal states delegate their spatial planning to the regions for regional spatial planning. The lowest hierarchy in line is municipal planning which provides exactly notable areas within the cities, defining, which buildings can be built in the zoning plan of the cities. At the lowest tier, the responsibility increasingly shifts from public building law and the creation of a land use plan to the private building law and the established development plan, generating a local zoning map. Within the government structure, spatial planning becomes more and more distinct by following the hierarchy and getting smaller scaled (Borchard, et al., 2011, pp. 567–570; Dammert, 2019, pp.108–110). Throughout the continuous development of legal conditions over time, the ordinance on the use of land for building purposes *Building Utilisation Ordinance (BauNVO)* was initially published in 1962 (Bau NVO, 2023). As part of spatial planning, it designates building land according to its utilisation in urban zones. Thereby new designation of industrial zones was prohibited within city centres and only permitted in outlying areas or specifically designated industrial areas, marked as *GI* on the zoning map. With reference to commercial real estate which has less impact on its surroundings, commercial areas, marked as *GE* on zoning maps, are listed separately and are even permitted within city centres (Wirth & Schneeweiß, 2023).

The idea of designating an area where enterprises can operate is shared by both industrial and commercial zones. The distinction between the two is that industrial areas are linked to more environmental issues, including air pollution, noise, steam, and bad smells. Furthermore, industrial zones are more dependent on connections to cargo infrastructure such as energy supply, heavy goods transportation, and freight transportation. Additionally, based on the industrial manufacturing in

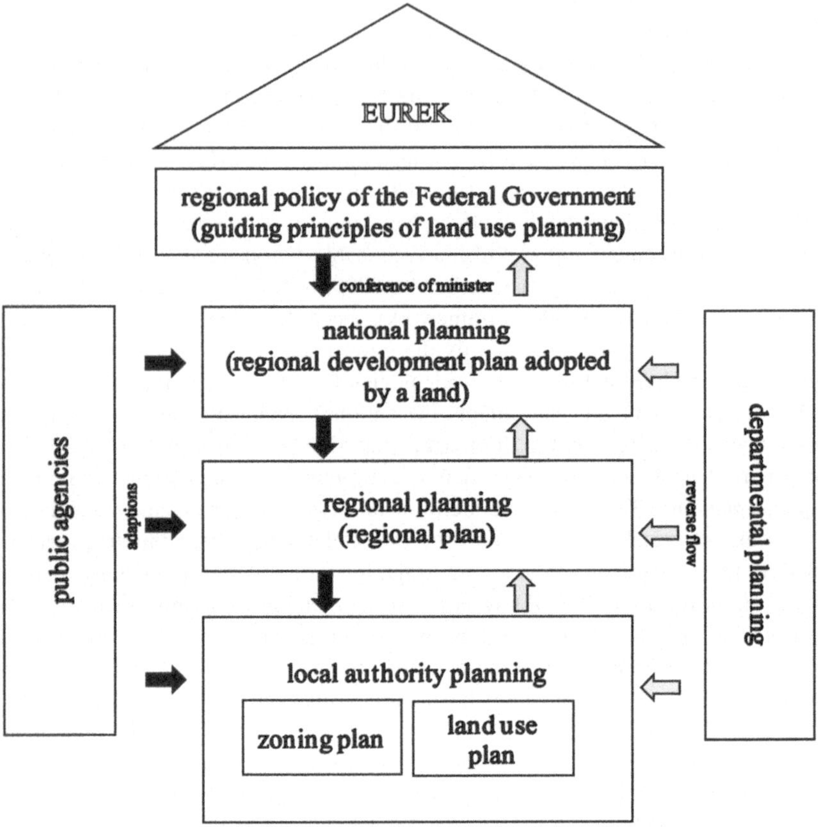

Fig. 2.3 Spatial planning system in Germany, own illustration according to (Borchard, et al., 2011, p. 568)

which they operate, they are subject to stricter regulations based on exposure (LogIVEST, 2024).

2.3.3.2 Building Protection

It is essential to consider the evolution since it has resulted in changes in attitudes toward historic preservation. Simultaneously, measures of preservation were undertaken, which had affected listed buildings. As the middle class became more educated in the nineteenth century, the idea of protecting historical buildings began to be discussed. Around 1900, the first approaches to heritage protection regulations focused on preserving an assumed ideal condition. At the beginning of the twentieth century, the previous work was perceived as a distortion of the heritage structures, and purification was carried out (Wild, Denkmalschutz Kompendium, 2023, pp. 4–5). Furthermore, George Dehio's perspective on Heidelberg Castle caused a significant new debate about conservation instead of restoration, which changed the way heritage sites were preserved from that time on. More precisely, it led to a shift from architectural design within conservation towards a greater focus on preservation (Wild, Denkmalschutz Kompendium, 2023, pp. 7–8). The argument between conservation and restoration in the early twentieth century ended abruptly when Germany's cities were extensively destroyed during *WWII*. The war and the Allied bombardment destroyed a huge number of buildings, artefacts, and the urban environment. Numerous heritage records were lost, as several buildings of the heritage protection authorities buildings were also destroyed. Cities have developed unique approaches to preservation due to the lack of a framework, which has resulted from post-war ambitions and temporary administrations that did not prioritise heritage conservation. Since there was no national legislation for the preservation of historical structures, local communities collaborated in order to maintain historic buildings in domestic regions and established their preservation criteria (Zwegers, 2022, pp. 28–29). This continued until 1951, when the heritage authorities of western Germany began to work together to set up a federal heritage conservators conference (Harzenetter et al., 2016, pp. 12–14). However, modern heritage protection only emerged in the last decades of the nineteenth century. Originating in art history, heritage preservation initially placed great importance on the preservation of castles, palaces, and churches and assigned the works to the corresponding epochal artistic styles (Föhl, no date, p. 2). In 1964, the *Venice Charter*, which emphasised the inseparable historical background of monuments, had a strong influence on heritage protection in Germany (Wild, Denkmalschutz Kompendium, 2023, p. 9). The federal government system reintroduced a decentralised political structure within the constitutional system, resulting in differing priorities for heritage preservation in the federal states. On the one hand, the heritage authorities developed guidelines, while on the other hand, the public was sceptical about heritage preservation caused by national extremism. During that period, the war caused massive problems which led to a shortage of resources, a lack of funds for investments, and multiple other problems that were more focused on than the preservation of heritage (Zwegers, 2022, p. 29). Due to federal states being responsible for heritage preservation, there is no national preservation law that is superior in the hierarchy towards the individual established. The goal of preservation thereby is to safeguard, maintain, and sustainably manage heritage sites. In the new utilisation, these sites must remain accessible to the public and scientific exploration (Wild, Denkmalschutz Kompendium, 2023, p. 12). In defining heritage and determining protection, a distinction is made between architectural heritage, garden heritage, ground heritage, movable heritage, and heritage areas (Wild, Denkmalschutz, 2024, p. 14). Heritage professionals furthermore distinguish between an intentional heritage, which can be linked to monuments like statues or triumphal arches, or a construction that became a monument due to historical events and the value assigned to it (Wild, Denkmalschutz, 2024, p. 4). Germany distinguishes between the terms *built heritage conservation, built heritage protection,* and *heritage doctrine. Built heritage conservation* (Denkmalpflege) indicates the practical application of construction preservation techniques. *Built heritage protection* (Denkmalschutz) specifies the legal-administrative protection procedure and authorisation of structural modifications. *Heritage doctrine* (Denkmalkunde) covers the area of scientific exploration, historical building research, and publications (Wild, Denkmalschutz Kompendium, 2023, pp. 2–3). The scope of built heritage includes cemeteries, monuments, mining or manufacturing facilities, military structures, sacred buildings, and pre-industrial and industrial structures (Harzenetter et al., 2016, p. 20). The scope of protection can apply to an entire building or only to specific sections of it. Furthermore, protection can be specifically applied to the interior, halls, gardens, or ensembles within the urban realm. Thus, the significance thereby is not attributed to a single building but rather to the appearance of the ensemble. Historic town centres, worker settlements, and industrial structures along rivers, including hammer mills and mills, are examples of topographical landscapes (Wild, Denkmalschutz Kompendium, 2023, pp. 129–134). The public's interest, which initiates the protection process, drives the protection effort. Accordingly, safeguarding cultural assets for future generations is one of the goals. Heritage preservation and conservation serve to highlight and strengthen the significance of the heritage that is under protection. As a result, it is necessary to prevent the damage and modifications that might be undertaken. The emphasis is on preserving the value by the correct care and appropriate use. In addition to the authorities, this responsibility is also given to the owners and all other relevant parties (Wild, Denkmalschutz, 2024, p. 7). When a building or site is protected, it must be listed in the publicly available heritage list by the federal state authorities and online (Wild, Denkmalschutz, 2024, p. 14). The listing includes a description of the entry and the justification of heritage status according to the heritage protection regulation, fundamental individual elements, and the scope of the heritage protection. The registration text should specify which distinct components are protected, whether it is the entire building, the outdoor area, the external appearance, or specific parts of the interior (Wild, Denkmalschutz Kompendium, 2023, pp. 121–127).

Germany's heritage authorities are not uniformly structured throughout each federal state. Most federal states have specialist authorities and heritage protection authorities that serve as higher-level entities or advisory offices. Thus, the management of protection at a local level is the responsibility of the so called 'lower heritage protection authority' in combination with local authorities. Due to their restricted area of influence, they supervise approving applications for amendments and preservation techniques simultaneously. For the state, fundamentally and nationally important concerns of protection, legal challenges, and law enforcement guidance fall within the responsibility of the so-called higher heritage authority. The specialist authorities are either independent consultants who provide appraisal, technical, scientific, or methodological assistance, or they are part of the higher heritage authority (Wild, Denkmalschutz Kompendium, 2023, pp. 97–100). If a modification of a protected building or site is planned, an approval procedure under the heritage law must be completed. The purpose is to prevent harmful changes while controlling adaptions on the buildings and supporting new uses. With this process, the authorities can review whether significant harm to the building would be caused. Authorisation should only be opposed by interests that carry greater weight and not for minor changes. In addition, authorisation must be granted if the public interest in alteration and conversion to a new use outweighs other issues (Wild, Denkmalschutz Kompendium, 2023, pp. 175–179). By mentioning principles of preservation, the central demands and ideals that have emerged from the regulations and the *Venice Charter* can be highlighted to fulfil an application for maintenance and modification of heritage construction. These are authenticity, preservation of context and appropriate environment, preservation through use and maintenance, reverence and proportionality, limitation to what is necessary, repair before renewal, material and work justice, reversibility, and distinguishability of old and new, preliminary scientific investigation, documentation and public access (Wild, Denkmalschutz Kompendium, 2023, pp. 27–39). Different systematic and technical approaches of protection thereby include methods of conservation, maintenance and repair, refurbishment, modernisation and renovation, restoration and exposure, copy, translocation, reconstruction, or graphic reconstruction. The appropriate approach and technical procedures to be applied must be determined on a case-by-case basis through consultation with the responsible local authorities (Wild, Denkmalschutz Kompendium, 2023, pp. 40–64; Wild, Denkmalschutz, 2024, 22–24). Besides the approval of heritage authorities with listed building consent, general planning permission might also be necessary in some refurbishment projects. Thereby the approval must be granted by the state building code of the federal states. If both applications are required, the building application typically includes the listed building planning permission. If approval for structural measures is granted, the exact measures applied for and approved must be carried out within two to four years, depending on the individual federal state. Authorisations often contain additional conditions for implementation. Authorities use these conditions to control the measures carried out, for example by binding them to offers submitted. It is frequently beneficial for the builder to adhere to conditions and requirements, as they may consequently generate tax advantages (Wild, Denkmalschutz, 2024, pp. 28–29). Benefits for taxation can be granted as special depreciation by the authorities for the necessary and meaningful costs allocated to the preservation of the heritage building. The tax benefits arise out of the Income Tax Act and must be applied for. The authority will review the application and determine whether measures were necessary for the preservation or not. Previous planning with the authorities and the fact that the measures were implemented by a specialised firm rather than internally are relevant for tax benefits to be granted (Wild, Denkmalschutz, 2024, pp. 48–50). Besides potential tax savings, there are additional financial support instruments available if local heritage authorities issue a confirmation of eligibility under heritage protection. State funds, municipal funding, subsidised loans by *KfW* Efficiency House Heritage Building "KfW-Effizienzhaus Denkmal", or bank loans could be sustained (Wild, Denkmalschutz Kompendium, 2023, p. 77). Furthermore, local funds by authorities might be available for urban developments such as facade design and urban beneficial development by creating social or green spaces. A main issue with financial claims often is the limited scope of applicants and that measures are not allowed to be started before funding is approved (Wild, Denkmalschutz Kompendium, 2023, pp. 77–78).

2.4 Number of Historical Industrial Buildings

2.4.1 United Kingdom

Given that the *Industrial Revolution* originated in the *UK*, it has had a significant influence on the prevailing industrial heritage. The first industrial buildings were constructed in the *UK*. These include the first multi-storey cotton mills due to the invention of the iron-framed building technology (Clarkson, 1990, p. 50). Its status as the world's leading country in the cotton and woollen industry has increased the number of production sites existing in the *UK*. London's role as a national port of call and the fact that a large quantity of goods were transported via the canals additionally led to the high number of warehouses, mills, and industrial sites along the canals (Ziegler et al., 2012, pp. 30–31). Although a long history of the industrial era exists, the introduction of heritage protection lasted long. Due to weak conservation legislation after its introduction and with progress in time and technical developments, numerous structures that would be designated as heritage today have already been lost (SPAB, no date). After the outbreak of *WWI* and aerial warfare, further losses were documented. This resulted from a focus on targeting production sites, to prevent the creation of military equipment (Lüken, 2014). Even more significant were the losses during the *WWII* in which air strikes were used more intensively (Fig. 2.4) (Taylor and Francis Group & The National Archives, 2019).

Determining an exact number of built heritage sites is challenging after each of the four countries undertakes listing in their responsibility and the total number changes annually due to new listings and delisting. Regarding official numbers by the ministry department *DCMS,* they indicate over half a million built heritage sites and approximately 10,000 conservation areas (Towell & Pugh, 2007, pp. 11, No. 6; 12, No. 8). The *NHLE,* which provides the list for *DCMS* currently records over 400,000 listed heritages (Historic England, About The List, no date). *HE* provides information on 45,000 industrial locations and industrial heritage constructions on their list (Historic England, Support Your Industrial Heritage, 2011). Furthermore, due to functional redundancy, many of them are at risk of disrepair. A survey by *English Heritage* revealed that the percentage of listed industrial buildings at risk is three times higher than the national average (English Heritage & bdrc continental, Industrial Heritage at Risk Public Attitudes Survey, 2011).

2.4.2 Germany

Referring *Fig. 2.5*, the map depicts the effects of bombing on German city centres, highlighting urban areas with over 40% destruction during *WWII*. Further demolition began after the war to tackle the shortage of resources and rebuild urgently required residential buildings (Rukschcio, et al., 2019, pp. 12–14). In general, there is no comprehensive overview of the destruction caused by the war and certainly no precise number of exact losses in the post-war decades. However, even during the post-war reconstruction process of cities, some citizens and associations appreciated historically significant buildings and areas and protected them (Kiesow et al., 1983, pp. 4–5). During that period, the National

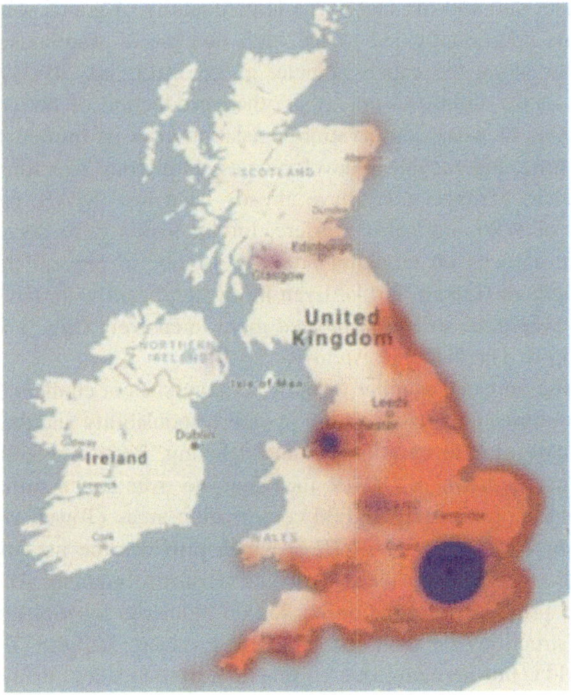

Fig. 2.4 "A heat map of the United Kingdom, 1939–1945. The Colour red shows the density of air raids and the colour purple represents numbers of casualties (injured and killed)" by (Taylor and Francis Group & The National Archives, 2019)

Committee of Heritage estimated the existence of half a million protected artefacts. Appreciation for heritage increased as the nineteenth century progressed into its later decades. The broadened understanding of built heritage no longer only included aesthetic historic buildings but also technical facilities, transportation constructions, industrial buildings, and industrial areas. All those buildings were valued for their association with constructions erected by their ancestors, combined with a uniqueness, artistic quality, and cultural-historical significance (Kiesow et al., 1983, pp. 4–5). Again, determining an accurate number of built heritages is challenging, because since 2012 such figures are no longer accessible at a single source anymore. Each federal state is responsible for calculating built heritage independently, leading to a lack of standardised counting procedures. However, experts estimate the total number to be around 1.3 million (Liersch &

Asef, 2018, p. 29). Over a million heritage sites were identified in 2017 through a data collecting compilation from all federal state authorities. Within a proportion of 63%, architectural heritage has the largest proportion of this number, including individual protected constructions, garden protections, and protected areas. In relation to the total number of constructions in Germany, a share between three to three and a half percent can be attributed to protected built heritage (Liersch & Asef, 2018, pp. 7, 29–30).

2.5 Risks and Trends for Historical Industrial Buildings

Facing the development of cities and urban areas in an international context it can be noted that they have undergone, and continue to undergo, permanent change over decades and centuries (King, 2015, pp. 10–11). These transformational processes thereby concern the adjustments of cities, urban areas, regional patterns, and national and international structures. Furthermore, cities are influenced by broader spatial, cultural, social, political, legal, technological, and many more micro-effective developments (Ewen, 2016, p. 1; Adler, 2018, pp. 10–12). With the dispersion of urban lifestyles, several trends of urban developments can be observed, such as urbanisation, suburbanisation, desuburbanisation (Turok & Mykhnenko, 2007, p. 166), and the recently established trend of reurbanisation (Fig. 2.6); (Dembski et al., 2021, pp. 2–3; Siedentop, 2008, pp. 194–95). Those four separate phases can be seen as important components of the gradual development of technology, economy, and social changes, which produced the many stages and movements within cities and metropolitan regions (Allen, 2009, pp. 116–119; Ewen, 2016, pp. 12–16).

It is important to point out the substantial influence of recent technological innovations on social, every day, and professional life both past and present. These universal changes have generated and will still be generating new demands for spatial planning within cities and metropolitan areas. New requirements must solve the simultaneous requirements for social places, labour, and housing (Graafland, Hauptmann, & Sassen, 2001, pp. 655–656). The rising demand for living in cities is further driven by so-called prevailing *megatrends* which are distinguished by duration, ubiquity, globality, and complexity. *Megatrends*, for instance, include urbanisation and globalisation, with sub-trends such as the 15-min city, urban manufacturing, condensed spaces, and globalisation. Additionally, there is a great number of different megatrends and sub-trends that should be considered in development scenarios (Zukunftsinstitut, 2023).

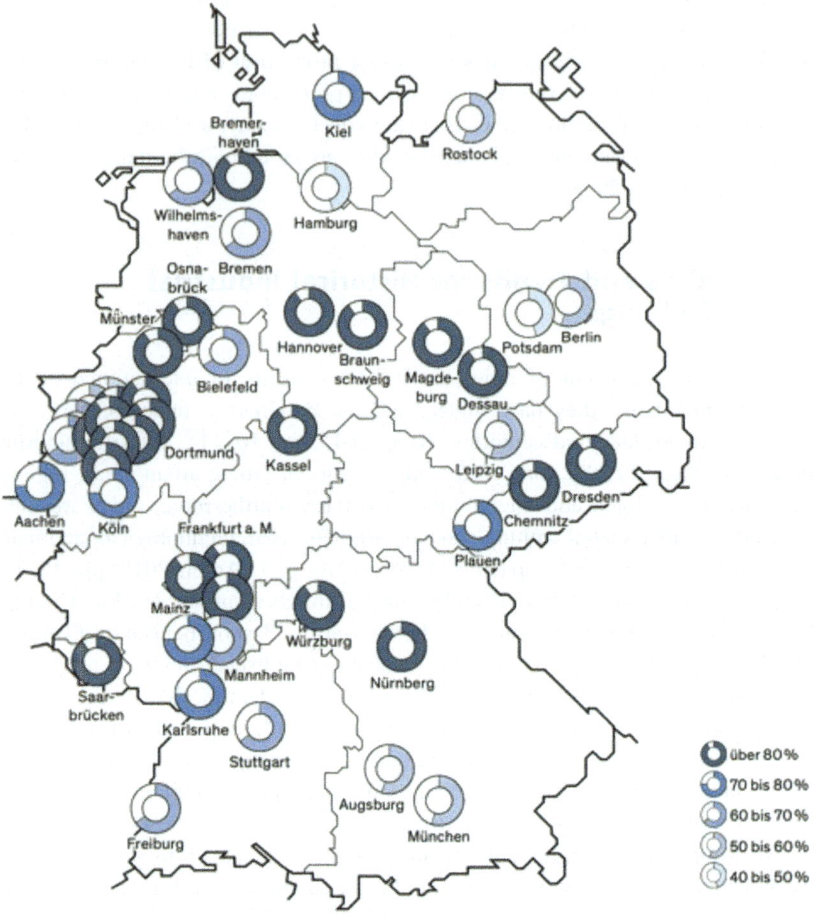

Fig. 2.5 Wartime destruction in large cities 1939–1945, City centres with a degree of residential destruction of over 40% Source: Volker Bode 2002 by (Rukschcio et al., 2019, p. 15)

This results in a new requirement for cities, their transformation, resilience, and urban development. Due to the need for agility and adaptability in cities, new concepts and models must be developed and alternatives to previous uses of already built-up land must be derived (Pfnür, Eberhardt, & Herr, Transformation

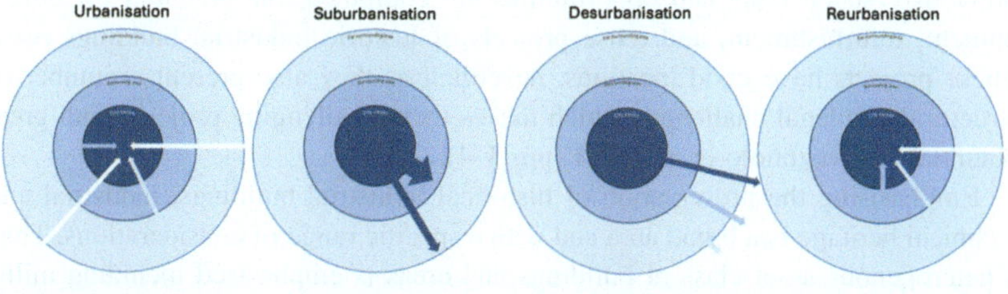

Fig. 2.6 Phase model of the urbanisation cycles of population distribution, own illustration according to (Ravin, 2017)

der Immobilienwirtschaft—Geschäftsmodelle, Strukturen, Prozesse und Produkte im Wandel, 2022, pp. 18–20). When it comes to the refurbishment of buildings, constructions, and sites a detailed examination of the property and the buildings situated on it should be undertaken. Expert surveyors might be considered to analyse the existing constructions according to, for instance, the building stock if there are risks, such as building materials containing asbestos, and construction waste or soil in the ground. In addition to standard investigations undertaken, regarding built heritage, the protection of parts of the building, the entire construction, or the surrounding area must be analysed to identify risks and potentially cost-intensive aspects. Regarding the development of listed buildings, the time factor of planning and receiving approvals from building and heritage protection authorities must be considered (Köster, 2021, pp. 38–39). An even more comprehensive assessment is required for the redevelopment of former industrial sites. Additional risks that might appear by dealing with the refurbishment of industrial sites might be animal biotopes, protected animals living on site, or landscape protection which must be considered. Furthermore, a problem that might arise is harmful soil with contaminated land since it had been previously used for industrial purposes. In Germany, there is an approximately number of 68,000 presumed former landfills and about 193,000 presumed sites are used for industrial purposes (Glatte, Location Strategies, 2024, pp. 100–102). Comparing it to the *UK,* it is estimated that approximately 325,000 potentially contaminated sites exist (The Reigate & Banstead Borough Council, no date). Besides the risk of contaminated land caused by careless industrial use or during the world wars by the destruction of industrial sites, there is an additional risk of unexploded ordinance lingering in the ground. Industrial complexes had been one of the prime targets of air raids and not only caused immense losses to industrial constructions but also caused additional harm to the sites (Zwegers, 2022, p. 135). When the previously stated risks and opportunities are combined, the potential for development, refurbishment, and reuse projects of historic industrial buildings rises. These projects have good locations, nevertheless, they also present a number of structural and legal challenges which makes them challenging projects with great potential (Gravagnuolo et al., 2021, pp. 3–4).

Emphasising the preservation of historical industrial buildings, industrial and technical heritage is a broad area and with a specific range of considerations. Thus a heterogenous asset class of buildings and areas is emphasised including mills, mines, factory buildings, waterworks, gas tanks, petrol stations, bridges, and all varieties of technical equipment for production is included. These industrial sites, factory buildings, and technical facilities from different centuries can be individual buildings, constructions with outdoor facilities, groups of buildings, or entire historical industrial areas (Wild, Denkmalschutz Kompendium, 2023, pp. 82–83). Industrial areas may furthermore include industrial residential centres, administrative buildings, or even transportation networks. Focussing on industrial heritage sites, they differ fundamentally from other asset classes such as residential buildings in terms of their construction background, use and wear and tear, fittings, and the potential for third-party usability. When it comes to this distinct asset class of protected industrial buildings and converting them into contemporary residential or office uses is often the only way to preserve them. Challenges arise in interior planning due to large industrial windows, which need to be optimised in order to produce the most usable space possible across multiple levels. Furthermore, lighting and thermal insulation are challenges due to being inadequate for modern living standards. Additionally, delicately designed roof structures are challenging to refurbish. When historic machinery has been preserved, it often leads to the prevention of new uses. Within the last years, developments of loft flats have become trendy, and industrial areas have been focussed by developers. Especially if broad road networks are present to develop as new sites and thereby increase the economic efficiency of projects. Besides that, large industrial warehouses were transformed into studios, or house-in-house solutions were incorporated. Nevertheless, ignorance of preservation might result in the loss of the protected heritage (Wild, Denkmalschutz Kompendium, 2023, pp. 82–83).

The refurbishment of historic industrial buildings is not just about cultural heritage and identity, it is also part of an urban development process and closely linked to urban transformation. One conflict exists is between the various stakeholders and authorities when it comes to preserving built heritage and allowing adoptions. A key question is how protection and preserving can go hand in hand. Key challenges that arise from the refurbishment are the gratitude of built fabric and character while decay is prevented and renewed vitality is integrated through sustainable refurbishment (Oeverman & Mieg, 2015, pp. 3–6).

2.6 Corporate Real Estate Management Versus Institutional Real Estate Management

In redeveloping former industrial sites, it is essential to consider not only from which period they originate, and current market trends but also their management history and the purpose they served for previous owners. Since the term industrial heritage or historical industrial buildings is used, the former operative use of the constructions is given.

Buildings were mostly owned by companies as an asset and have a significant share within the property market. The first professional approaches towards property management by companies emerged in the Anglo-Saxon area in the 1980s with the established term *Corporate Real Estate Management (CREM)*. This expanded to continental Europe during the 1990s (Glatte, Corporate Real Estate Management, 2021, p. 1). The definition evolved in Britain by R. Bon states that "'CREM concerns the management of buildings and parcels of land at the disposal of private and public organisations which are not primarily in the real estate business. An organisation which occupies space is in the real estate business and needs to manage it properly.'" (Haynes et al., 2017, p. 4). The definition of *CREM*, which is considered an integral part of the company, is based on services being provided for the main business purpose. *CREM* assists and supports the corporation in pursuing its business objectives and is managed comparable to other departments within the company (Haynes et al., 2017, p. 4). The German definition associated with *CREM* focuses on the cultivation, management, and usage of the corporate sites within the corporate strategy. It refers to properties implementing and supporting the core activities of the companies. It should thereby help to create transparency over costs by managing sites. Furthermore, the task is pooling the expertise for real estate to carry out site management derived from corporate strategy (Pfnür, Modernes Immobilienmanagement, 2011, p. 165). *CREM* is thus subordinate to the primary business activities. The main task is to provide support, and efficiency as a service provider for the major business area of the corporation (Fig. 2.7); (Glatte, Entwicklung betrieblicher Immobilien, 2014, pp. 13–14).

Since industrial heritage sites do not fulfil their industrial use any longer and after they have been decommissioned, a fundamental change must be implemented. The change not only involves structural alterations on the property itself

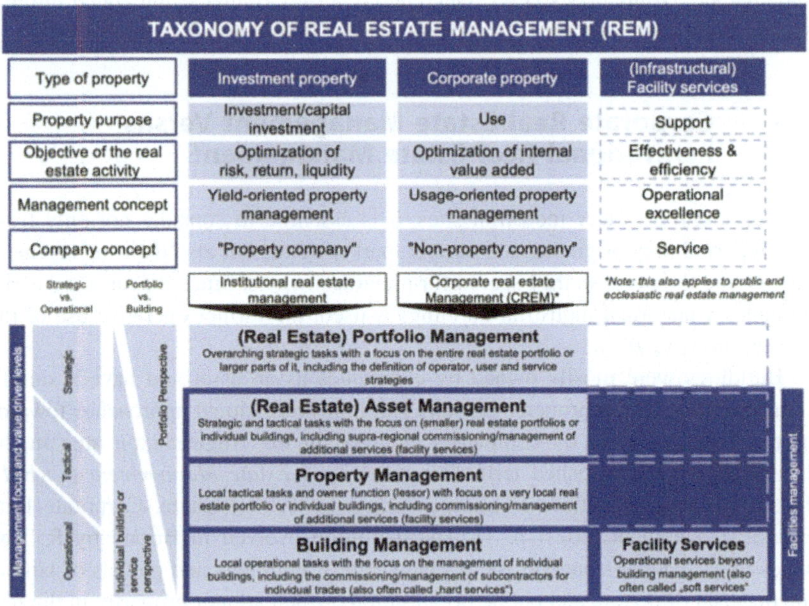

Fig. 2.7 Taxonomy of Real Estate Management REM (Glatte, Location Strategies, 2024, p. 7)

but furthermore by a change in the management system. Previously, the property was managed as a corporate property under *CREM* driven by a service character. The refurbishment process marks the transition towards institutional real estate management under the ownership of a property company. Derived from the fundamental change, the new management approach is based on an investment concept focussed on optimising risk, return, and liquidity. Accordingly, management is now yield-orientated (Fig. 2.7); (Glatte, Entwicklung betrieblicher Immobilien, 2014, pp. 297–298; Glatte, Location Strategies, 2024, p.7).

Shifting the perspective from a *CREM* to an *Institutional Real Estate Management* also alters the approach to property and building valuation. The shift simultaneously indicates that industrial properties previously owned by corporations have emerged as an asset class within the investment landscape. A property that is included in the company's balance sheet has often been valued by the property/asset value method (Sachwertverfahren) in Germany. Through the change of perspective, the value of the property now needs to be derived from an income approach (Ertragswertverfahren), such as the capitalised earnings value method or Discounted Cash-Flow *(DCF)*. The comparative value method (Vergleichswertverfahren) furthermore is somehow irrelevant due to a lack of comparable properties and transactions within the market to derive meaningful figures (Leopoldsberger et al., 2022, pp. 143–145). Compared to other real estate asset classes, a unique characteristic of industrial properties is their limited third-party usability. This is caused by the intended construction process for a distinct use of the company such as a manufacturing process in a specific process chain. Thereafter redevelopment might require cost-intense measures (Leopoldsberger et al., 2022, pp. 147–148). Besides considering third-party usability, a second influencing factor is age. Due to their historical background and protected building fabric, industrial sites cannot easily be compared to other existing assets in terms of possible flexibility or energy efficiency, for instance. In line with the history and the former industrial use of the buildings, periods of vacancy present additional unique challenges. Contamination in the building fabric or soil are a few instances of what might need to be considered during the valuation process (Meister & Schulz, 2018, p. 756). In contrast to the Anglo-Saxon valuation system, the German approach values land and construction separately. The reason for this is based on the separate contemplation and possible usage of the site with a different construction. Caused of the protection of heritage and the duty to preserve the buildings, there sometimes is an assumption of a theoretically infinite lifetime. Although the protection allows one to assume an infinite life span, the useful economic lifetime differs according to possible tenancy. Thus, economic life depends on spatial structures and flexible utilisation options. Accordingly, with an obsolete original utilisation of industrial built heritage, the new value arises in the gross profit ("Rohertrag") of the construction (Society of Property Researchers Germany, 2007, pp. 21, 25). It is relevant for the valuation of industrial heritage sites or buildings to create a concept of utilisation. A master plan fulfils the intensive examination of the architectural heritage, the development of a utilisation concept for the heritage, and the creation of possible usable areas that can subsequently be rented out. In order to develop a master plan, experts should be consulted, and necessary surveys should be prepared. Through the process of a preliminary investigation, planning conception, and, if necessary, visual elaboration, a concept can be presented that will be necessary for the valuation process as well as for the building law relating to county policy (Society of Property Researchers Germany, 2007, pp. 17–19). Besides the risks previously mentioned, heritage sites and buildings can be desirable due to their remarkable architecture and being situated in locations that represent potential. The existing challenges require an overall concept for planning in advance and a positioning of the existing areas with an awareness of the structures and flexible design of the existing areas by, for example, a large number of commercial users (Meister & Schulz, 2018, pp. 756–757).

In the *UK* there are three approaches to property valuation which are Market Approach Methods, Cost Approach Methods, and Income Approach Methods (Morri & Benedetto, 2019, pp. 47–48). Thereby sales comparison, income capitalisation, and replacement cost are valuation methods all based on the concept of comparison. The sales comparison approach estimates property value by examining sales of like properties. Income capitalisation is undertaken using a yield or goal rate of return, including possible net income usually derived from rent. The replacement cost approach projects the likelihood of creating a comparable replacement for the purchased property. These methods can be applied to frequently occurring types of property on the market but are not entirely suitable for specialised properties. Furthermore, in the *UK* the profis method and residual method are used. The profit method is the primary method used to value specialised properties due to the heterogeneity of these properties and the limited number of transactions. Although the Replacement Cost Method is frequently used for highly individualised sites such as industrial, it is not applicable for heritage sites due to the protection and their unique character (Wyatt, 2013, pp. 101–102, 171–172). The investment approach is employed to value properties that are retained as investments and are undergoing redevelopment. Planned cashflow is the income used for return on the investment which is either calculated with the *ARY*-method or *DCF*-method (Wyatt, 2013, pp. 115–118, 135–16).

Besides the traditional valuation methods, *Historic England* states that many other benefits arise from monuments and built heritage and that these often have no measurable market value. As a result, the *Total Economic Value* (TEV) approach also attempts to attribute further benefits. A distinction is made between three different types. These are the direct, indirect, and non-use values. These include the active visit to heritage sites, economic benefits from the visits through consumption, for example, and the monument or built heritage as a source of identity and relevance for several generations. Hedonic pricing is used here to quantifiably incorporate visitor flows and the impact on services and property prices (Historic England, Techniques for Valuing Heritage in Economic Terms, 2023).

Research Design 3

3.1 Research Inquiry and Hypothesis

Cities and the urban realm are continuously in a transformational process with a variety of heterogeneous issues influencing them. This is due to the limited availability of undeveloped sites within cities which poses difficulties to urban development. The consequence is the requirement for a wide range of creative approaches that commonly cover the multi-layered demands of today's world (UN-Habitat, 2022, pp. 3–4). From an international perspective, urban development is characterised by the *Industrial Revolution* initiated during the eighteenth century and spreading internationally. Cities, urban environments, and areas characterised by manufacturing in Germany and the *UK* possess architectural traces of history (Allen, 2009, pp. 2–3). Linked to history, formerly industrial buildings or entire industrial areas, some of which are located in urban A-locations, suburban B-locations, or surrounding regions are under preservation order in both the *UK* and Germany (Pfnür, Eberhardt, & Herr, Transformation der Immobilienwirtschaft—Geschäftsmodelle, Strukturen, Prozesse und Produkte im Wandel, 2022, pp. VII, 40). Both in the urban realm and peripheral locations, abounded sites are famous as lost-places. The colloquialism name implies that the former use is no longer practised, and the buildings have been partially left to their fate and external influences. Urban developments and project developments are continuously challenged by newly emerging hurdles under globalisation, reurbanisation, and climate change. Since both urban and more rural environments feature historical industrial buildings, the question arises as to how these contemporary constructions might be repurposed. The issue arises from revitalising the existing buildings of listed sites and integrating them into the property market, cities, and social structures. Due to historical similarities, an international comparison is to

R. Groer et al., *Historical Industrial Buildings and their Real Estate Utilisation*,
Studien zum nachhaltigen Bauen und Wirtschaften,
https://doi.org/10.1007/978-3-031-89417-6_3

be made in this study in order to compare the regulatory framework, the handling, and the revitalisation of historical industrial buildings. The paper's core research question is which distinctions exist between Germany and the United Kingdom in their approaches to handling historical industrial buildings and potential developments. The sub-questions arise if strategies and possibilities for redevelopment can be determined from an international comparison.

3.2 Methodology of Research

In order to analyse the research question, the research design was chosen to collect data from qualitative expert interviews. The qualitative methodology is used to summarise and explain the results of the interviews in words. The empirical method of the thesis is based on the structure illustrated in Fig. 3.1 as the ideal approach.

The research question concerns a multi-layered and complex issue, which includes the subjects of heritage protection, industrial buildings and areas in the context of historical development up to the present day, and revitalisation, which is associated with real estate project developments. In order to understand all these issues in a common context, to merge, and to explain them thematically, a theoretical preliminary study was primarily conducted. The in-depth analysis was the basis of preparatory work to be able to engage with professionals in an expert-interview on the specialised field, with only a restricted number of occurring cases. In addition, the thorough preliminary work with the various heterogeneous subject areas served to prepare the basis for the subsequent creation of an interview guide (Gläser & Laudel, 2009, pp. 107–108).

Experts from various professional groups were requested to approach the topic of revitalising historic industrial buildings from a broad spectrum and to obtain a comprehensive overview of the subject area in an international comparison. Accordingly, the experts, based on different professions and thus activities within the revitalisation process were chosen to represent different viewpoints. The clustered search for experts was conducted of project development, construction law, architecture, and additional advisory functions. Due to the special real estate asset class and a limited number of cases in both countries, the research for experts and arranging interview appointments presented a challenge. The experts were chosen by conducting a search for projects, related to the revitalisation of historical industrial buildings in a variety of sources, including online, in books, and within reports. Thereafter sixteen German and nineteen English candidates were contacted in order to win them as interviewees. Four interview partners agreed

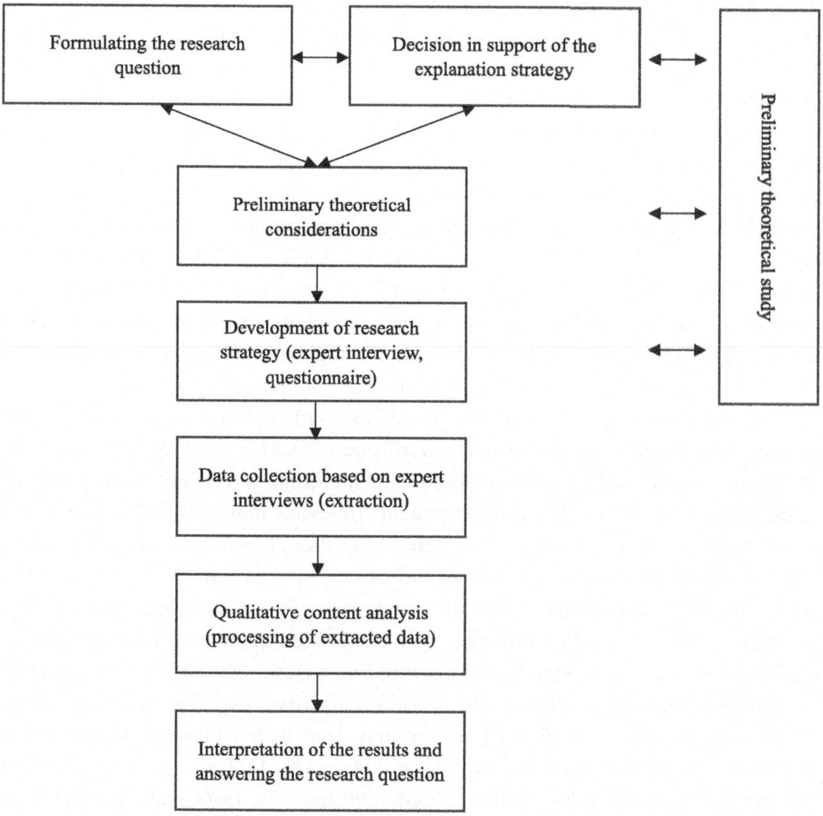

Fig. 3.1 Empirical method. own illustration according to Gläser / Laudel (Gläser & Laudel, 2009, p. 38)

to be interviewed on the German side and five on the English side based on the inquiries and phone calls obtained. Unfortunately, two architects from Germany replied too late to arrange an interview appointment and could therefore not be included in the study. Due to the physical distance within Germany and particularly to the *UK*, all interviews were conducted equally online via Zoom (Gläser & Laudel, 2009, pp. 117–119). Regarding different professions and levels of experience in the field, a semi-standardised interview guide was considered appropriate with regard to the required data collection (Gläser & Laudel, 2009, p. 41). The interviews took place between the 29th of June and the 5th of July

2024 and lasted an average of 65 min. With the consent of the interviewees, the interviews were recorded for the subsequent transcription. The recorded Zoom calls were transcribed using principles, anonymisation, and smoothening. Names of the interviewees were anonymised, as well as information provided by the interviewees on companies, colleagues, clients, and projects that could be used to draw conclusions about the person or company. In addition, the identities of cities and regions were anonymised. All anonymised contributions of the records were distinguished with a letter abbreviation, an abbreviation of the country name, a number of the chronological order of the interviews, and a numerical classification system. This implies that, for instance, the number of different projects mentioned can be identified, however, the immediate identification of the interviewees by this system can be pretended, although unfortunately, it might not be impossible by more thorough research into the spoken dialogue.

The transcripts were created in a word-for-word format. Language and punctuation were slightly smoothed out and affirmative and confirmatory sounds such as "mhm", "aah", and "yeah" insignificant in terms of content were removed. In addition, to improve readability, content-irrelevant individual words that were begun multiple times during, for instance, the thought process were also eliminated. Each speaker's contribution was identified by a paragraph, the anonymised abbreviation of the speaker, and the time when the speaker changed (Fuß & Karbach, 2019, p. 30). Accordingly, the basis for the qualitative research method used here are the transcribed semi-structured interviews, which originate from the recorded verbal data. Analysing the spoken transcribed expert interviews, it must be pointed out that a qualitative survey may lead to inaccuracies in the results due to the chosen research design (Gläser & Laudel, 2009, pp. 43–44). The analysis was carried out using qualitative content analysis. Data was systematically extracted from the transcripts and thus presented. The extracted data was then used to answer the research question and to show differences between the two countries by comparing them. The information was extracted by defined categories. The categories were developed from the theoretical principles and derived from the influencing factors in the context of the research question. The category system serves as a basis but is not a fixed construct as in other methods, allowing further categories to emerge as information is extracted from the transcripts. The categories are focused on the relevance and influencing factors of heritage properties. In addition, there are the categories of heritage protection and planning process, approaches and experiences with revitalisation strategies, financial grants, problem areas, and prospective trends for revitalisation. By extracting the data, it can be summarised, processed, and accordingly is available for the final step of interpreting and discussing the collected statements (Gläser & Laudel, 2009, pp. 199–204).

3.3 Design and Execution of the Questionnaire

Conducting expert interviews with the selected research design of semi-structured interviews requires an interview guide that has been prepared in advance. The pre-formulated questions served as a framework for the interview without, however, specifying a procedure for the questions or a standardised, scheduled structure. The questions aimed to inquire about the subject matter of the influencing variables for the research question and generate statements about the everyday practical relevance of the experts. The chosen structure ensures that statements are generated that can be compared afterward. The questions were prepared open-ended to encourage the interviewees to share their knowledge, experiences, and perspectives. In order to bring the interviews closer to a natural conversation format, thematic categories were formed, which were derived from the literature research and the examination of exemplary projects of revitalisation of historical industrial buildings (Gläser & Laudel, 2009, pp. 142–150). The conducted interviews started with a request for permission to record. Furthermore, questions about the interviewee's background, expertise, and connection to the revitalisation of historic industrial buildings were asked. The first thematic category of the questionnaire addressed the importance of historical industrial buildings. The experts were inquired about their perspective on the asset class and the relevance of this sub-segment of the real estate industry. Besides that, the challenges and opportunities in the subject area for revitalisation and new uses were inquired about. The introduction concluded with the question of the scope of the historical industrial property asset class in terms of single buildings or industrial areas. The second category concerned questions about heritage protection. This involved queries about the scope of protection, restrictions on construction measures, and simplified procedures for conservation. It was additionally asked how the preservation of the historic fabric can be achieved while meeting the challenge of altering the buildings under modern requirements. The third category of the questionnaire addressed the planning process in relation to legal and construction challenges and which stakeholders are involved in the process of revitalisation of the built heritage. The fourth category enquired about financing options, particularly with a focus on possible subsidies available from the government, or other financial support institutions. The fifth category dealt with the subject area of revitalisation and the resulting problem existing within the process. Questions were asked about the appearance and existence of environmental or man-made problems and their impact on the existing building stock. Furthermore, questions were addressed regarding structural changes and cooperation with the authorities. In connection with the authorities' planning and revitalisation, the economic feasibility of such projects and possible uses were questioned. The sixth and final category concerned questions of existing trends within conversion and different revitalisation concepts and approaches. The influence of *ESG*-factors and the revitalisation of buildings and areas in the scope of urban development was inquired about. The interviews were concluded with inquiries regarding additional influencing factors and aspects of the structures' utilisation in the real estate sector.

Results Presentation from the Interviews

4

This section presents the results of the content analysis according to *Gläser and Laudel*, which were extracted from the transcript of the semi-structured interviews. The categories already mentioned are used to present the results thematically categorised.

4.1 Overview of Interviewees

4.1.1 United Kingdom

The United Kingdom's perspective was achieved through interviews with five participants from a variety of professional groups and business sectors.

- RUK1 is a Joint Managing Director with 15 years of professional experience in the letting and operation of real estate at a family-owned development company.
- RUK2 is a legal counsellor, specialised within the real estate sector on use changes and developments of properties with 20 years of professional experience.
- RUK3 is a director of building redevelopment in an engineering consultancy with 14 years of professional experience and is furthermore conservation accredited.
- RUK4 is a partner of an architecture office with heritage consultancies with 40 years of professional experience.

R. Groer et al., *Historical Industrial Buildings and their Real Estate Utilisation*,
Studien zum nachhaltigen Bauen und Wirtschaften,
https://doi.org/10.1007/978-3-031-89417-6_4

- RUK5 is a director of an architecture office specialised in refurbishment projects including heritage with 25 years of professional experience.

4.1.2 Germany

The German perspective was achieved through interviews with four participants from different professions and business sectors.

- RG1 is a solicitor specialised in real estate law consulting for project developers with 25 years of professional experience.
- RG2 is head of valuation in a bank with 20 years of professional experience for several asset classes within the real estate industry.
- RG3 is a managing director in the area of valuation advisory with 27 years of professional experience.
- RG4 is a spokesperson of a board with 30 years of professional experience within the Corporate Real Estate and urban district development.

4.2 Importance of Historical Industrial Buildings

The first section of the questionnaire addressed the classification and relevance of the redevelopment of historical industrial buildings and their significance for the real estate industry. Challenges and opportunities existing and the scope of buildings in relation to the answers given by the experts were requested.

4.2.1 United Kingdom

Regarding the introductory question, the following answer from an expert provides a broad perspective on the subject. "So what we find with a lot of those is that they are incapable of being reused for original purposes. So very few of them are reused for industrial purposes, but they have a very high social value and interest in terms of architecture and they are very solid, reliable buildings. So what we tend to see is them being reused for other purposes, especially residential. There is a lot of desirability in having repurposed industrial buildings for residential, but also increasingly office floor spaces and creative and artistic floor space. So that with it brings kind of like a change of use in the type of areas where they are." (Interviewee, C.1.2, 1:18). Primarily, all *UK* experts answered that the buildings are no longer suitable for their original use and need to be repurposed, and therefore a mix of different uses as a strategy is required. Stating that these projects are highly complex with a variety of different issues and that they are cost-intensive projects. The interviewees stated that the constructions bundle social, historical, and architectural values and thereby attract numerous individuals who associate them with being desirable. In the context of new utilisation opportunities, all experts addressed commercial use combined with residential (RUK1 et al.). One interviewee highlighted the demand and fashionable appeal of residential use in the redeveloped areas (RUK4). Nevertheless, also challenges have been stated regarding residential use including the spatial structure and technical obstacles, referring to the implementation of kitchen and sanitary rooms with necessary pipelines and costs arising (RUK5). Another expert mentioned noise, barrier-free access, and parking as restrictions for residential reuse (RUK2). All experts emphasised the great importance of urban development and placemaking in the course of redevelopments and thus new emerging use concepts being created (RUK1 et al.). The buildings' and sites' attractive locations and their economic and social significance were seen as catalysts associating them with people-centred environments (RUK3, RUK4). All experts emphasised that the local authorities have a decisive influence and the authorities themselves sometimes drive the developments. Besides that, all interviewees mentioned that refurbishments of historical industrial sites accompanied sustainability by preventing demolition and preserving valuable and robust structures (RUK1 et al.). One expert particularly emphasised that several sustainable practices were and still can be adopted from historical buildings and stated them as architectural role models (RUK2). One expert pointed out that these buildings are structurally advantageous due to their robust creation and high-bearing capacity. Initially seen as unrealised potentials since 1990, they are suitable for various uses and allow alterations to the structural substance (RUK3). One respondent mentioned the term 'creative placemaking' while „use[ing] elements of culture, elements of heritage, and elements of commerce blended with the regeneration of the buildings and the opening up of green spaces, all working together in a balanced way to affect the successful regeneration" (Interviewee, C1.1, 2:03). Regarding the real estate industry one interviewee mentioned that such developments mostly occur if the market is already experiencing a shift from former industrial to new uses. Adding that these developments indicate a buoyant and high-value market when such developments are considered and initiated. Stating that furthermore, more creativity is required for redevelopments as land values are very high. One result of this is that once these buildings are developed, they usually draw attention from tenants willing to pay a premium for the properties (RUK2). One expert stated the development of buildings formerly used for industrial purposes is a growing niche driven by adaptive re-use, which, however, requires intensive investigation and examination of the buildings and their historical and structural background (RUK4). Facing the challenges of the redevelopment, all professionals agreed on regulatory and legal challenges arising within historical industrial buildings and their protection whether situated in a conservation area or designated a listed building. Indicating that the restrictions are greater for a listed building particularly grade one listed than for grade two listed ones. According to all experts, cooperation with local authorities and *Historic England* is required as well as extra listed building consent for alterations (RUK1 et al.). In the matter of project development challenges, three experts cited common problems with structural features that do not meet modern requirements due to the existing building stock and need to be adapted (RUK1, RUK2, RUK5). One professional mentioned varying floor-to-ceiling heights according to historical uses and the different levels of buildings as challenging (RUK2). Another expert stated that different triangular and squared floor plans differentiate in being wide or narrow due to the former industrial use and lack of modern amenities such as toilets and proper access and escape routes (RUK1). Two experts furthermore mentioned challenges about leaking roofs, highlighting historical north light roofs in mills as posing challenges (RUK1, RUK5). One expert added deep footprints, lack of natural daylight, and ventilation as further issues (RUK5). Three specialists claimed problems because of leaking roofs and thus water entering the inside of the buildings rotting the wooden interior (RUK1, RUK2, RUK5). Facing opportunities all experts agreed on unique character and historical value (RUK1 et al.). The buildings were associated with great character, making them attractive for certain types of uses and tenants (RUK1), high marketability resulting from great cultural and social value, acting as 'cultural anchors' and being able to enhance community identity and place-making (RUK3), and appealing for some occupants due to large windows, open spaces, and historical character (RUK5). Furthermore, good and likely flexibility and adaptability were mentioned as the 'cart horse' concept, describing the flexibility and resilience of historic buildings compared to modern, purpose-built structures (RUK3). Regarding the scope, all experts stated that every case must be taken into account separately, but due to previous industrial use, these are typically larger sites rather than just single smaller buildings (RUK1 et al.).

4.2.2 Germany

Initially, three experts mentioned the sites' locations and their resulting importance for cities and urban planning. They stated often-appealing inner-city locations of historic industrial buildings created over time out of urban growth and thus being very much of interest for conversion and redevelopment based on their locations (RG1, RG2, RG4). The importance of the buildings was mentioned by the experts, which results on the one hand from their history, but on the other hand from their aesthetic and architecture, which was emphasised while constructing these buildings. Thereby the interviewees mentioned the potential of these buildings and sites for cities and urban environments, adding that these properties are opportunities in terms of neighbourhood /district developments (RG1 et al.). Two interviewees both stated the development of such buildings in the field of industrial sites is limited and a change of use is almost always necessary, due to the original use being inappropriate and obsolete (RG3, RG4). Furthermore, both mentioned that the combination of energy-efficient renovations and strict heritage protection regulations make it difficult to reuse them and that innovative concepts are hindered by complex and contradictory regulations (RG3, RG4). All experts mention that the local authorities play a decisive role and that most authority acts differently concerning refurbishment projects and enquiries (RG1 et al.). Two experts highlighted the relevance of close cooperation with the authorities since they might initiate redevelopments themselves providing sites and continuously controlling the process through legal regulations (RG3, RG4). Regarding the challenges, two experts underlined the originally custom-built background of the buildings and consequently mentioned challenging conditions arising for conversion to new uses (RG3, RG4). Nevertheless, one interviewee mentioned the structural features of these buildings due to their solid and robust construction, high ceiling heights, and reasonable load-bearing capacity, which makes them attractive and suitable for conversion for integrating different uses (RG4). One expert stated that these properties are often crystallisation points for further developments and the historical reference is often appreciated by people through charming and atmospheric alternatives compared to monotonous new projects (RG3). All experts mentioned residential use and multi-use concepts with commercial tenant structures as possible utilisation concepts (RG1 et al.). Thus, the use as commercial zones was mentioned by all experts, especially regarding the interest of large-scale companies and investment institutions with the creation of tenant mix structures consisting of offices, services, and shopping facilities (RG1, RG3). Two interviewees emphasised the trend and at the same time the charm of developing loft flats in areas formerly used for industrial purposes (RG2,

RG3). One expert described the combination of experience and refurbishment as a significant momentum, in the context of integrating sustainability by preventing gratuitous demolition and new construction while developing cities through sustainable, innovative, and mixed utilisation concepts for those areas (RG4). All experts highlighted in addition to the location, aspects of architectural uniqueness and history. Robust building structures are appreciated by all experts as an advantage for transformations in connection with often large sites, which are of great importance for the city and neighbourhood development (RG1, et al.). Regarding the challenges, all experts highlighted the protection of historical industrial buildings, which involves legal obstacles and is a restriction regarding possible alterations to building structures (RG1 et al.). Asking about the relevance, three experts stated that the frequency is limited and that a low number of units prevails within the real estate industry but cases occurring are mostly large areas (RG2, RG3, RG4). According to all experts, the properties under examination in the asset class are unique, individual cases that demand thorough case-by-case analysis. This relates both to the sites that occur with special cut and equipped shapes and to the individual buildings situated on the properties. In the given examples, all experts referred to industrial sites and industrial areas with unique and various building types but are usually developed and brought to the market as an overall project. Furthermore, all respondents mentioned that the relevance lies in the preservation of important historical structures and that the cities and urban areas can benefit from the redevelopment of historical industrial sites (RG1 et al.).

4.3 Heritage Protection

The second category asked participants about challenges and opportunities resulting from heritage protection and the heritage system. It faced the preservation of heritage and, given that there are basic procedures in place within the system, it was asked how refurbishment and conservation could be achieved simultaneously.

4.3.1 United Kingdom

Initially, most experts explained the system of heritage protection with the distinction between the categories *Grade I*, *Grade II**, and *Grade II*. Mentioning the differentiation between the categories with an increasing level of protection and restrictions towards *Grade I* when it comes to handling them (RUK2, RUK3,

RUK4, RUK5). Additionally, the interviewees explained the system of the conservation areas, which has an area-wide impact on regions and incorporates all buildings within the conservation area being protected, focussing on the external appearance. All interviewees ranked conversion areas as the lowest tier of protection, nevertheless, the approval of the local authorities is required for external structural changes within the area and for all other kinds of planned alterations on buildings from this stage onwards (RUK1 et al.). One expert mentioned that even buildings that are not listed themselves, but are situated next to a listed building, also outside the conservation areas, are subject to restrictions on structural alterations so as not to harm the neighbouring building (RUK2). One interviewee mentioned the current discussion of introducing a *Grade III* listing for all existing buildings. He rated it as controversial as it could undermine the importance of higher-grade listings and the entire listing system. Furthermore, he added that almost all industrial assets benefit from not being *Grade I* listed, allowing more room for negotiation and adaptation to meet modern requirements, such as sustainability (RUK3). Furthermore, all experts mentioned that the *Listed Building Consent* (LBC) is the consequence of the buildings being protected which is an approval process if any alterations are planned (RUK1 et al.). The interviewee especially highlighted the required consent for changes to avoid criminal offenses (RUK1, RUK4). Two interviewees noted that although there might be harmful impacts on a listed building, still planning permission might be granted if the public benefits from the *LBC* are so great that they outweigh the damage caused by alterations to the listed building (RUK2, RUK4). The consents thereby needed to be applied for by local authorities which, besides preservation officers and *Historic England* as a statutory body, all experts mentioned as stakeholders within the process (RUK1 et al.). Three professionals stated that at the outset a thorough investigation of the building stock and history must be conducted in order to understand which parts are historically, especially of interest, which are of interest, and possibly some modern irrelevant parts, thus enabling the balancing act between redevelopment and preservation of the built heritage (RUK2, RUK3, RUK4). Collaborating with local authorities from the outset is essential, according to all experts (RUK1 et al.). One interviewee stated that sometimes famous architects disregard the urban environment while creating artistic formations, consequently, they do not always equal successful redevelopments of historical industrial buildings (RUK2). All experts mentioned that it always is a case-to-case consideration whether conservation or refurbishment is the best scenario and that there are no general rules for such refurbishment projects. Nevertheless, all experts highlighted that the best way of preservation is to manage historical industrial buildings and assign them a new use concept (RUK1 et al.).

4.3.2 Germany

Concerning heritage protection, two interviewees mentioned the different scopes of built heritage protection where either there is an area protection existing in which the external appearance is protected and buildings have to complement, only the façade of a building is protected, the whole building is under protection, or up to very compartmentalised and detailed protection of individual halls, rooms, and architectural design elements of the interior. Both interviewees stated that this is generally the decision of the local authorities, and that small-scale protection often poses major problems (RG1, RG2). One interviewee furthermore mentioned possible archaeological protection after excavations (RG1). Regarding cooperation with authorities, three experts replied that working with the heritage protection authorities is always an additional challenge that is very difficult, time-consuming, and thus cost-intensive (RG1, RG3, RG4). Every expert mentioned that the heritage protection authorities have extremely differing viewpoints and approaches in terms of protection and allowing preservation, precise restoration, refurbishment, or revitalisation in connection with the delimitation of new components (RG1 et al.). Two interviewees highlighted that it is continuously necessary to accept major compromises between the developer and the heritage authorities (RG2, RG4). One interviewee stated that investors prefer the development of sites without heritage protection in order to avoid problems with authorities during the planning and development process and thus avoid historical buildings (RG1). To undertake a successful redevelopment of historical industrial sites, three experts mentioned innovative concepts are required to achieve the balancing act between preservation and sustainable refurbishment (RG2, RG3, RG4). Two interviewees highlighted that successful redevelopment is achieved by experienced specialists in the field of heritage protection combined with architects who create a reasonable, sustainable concept from the outset (RG3, RG4). One expert mentioned room-in-room solutions sometimes being the only approach to successful redevelopments (RG4), while one expert highlighted the façade protection concept being the most reasonable and promising approach to the preservation of external appearance and buildings meeting modern standards constructed attached to the historically relevant facades (RG3). One interviewee noted that some investors even no longer participate in particular areas due to heritage authorities handling preservation with a strong subjective perspective and not allowing minor alterations. Furthermore, it was stated that the hierarchical lower heritage protection authority was sometimes bypassed due to the upper heritage protection authority within the federal state system having a more pragmatic and revitalising approach (RG4). After the experts mentioned many obstacles, they called for more openness and a more practical approach from the authorities in dealing with the revitalisation of listed buildings (RG1, RG3, RG4). One interviewee thereby stated that according to his professional experience, the *UK* is more open-minded in addressing revitalisation projects and has allowed more brownfield developments in the past (RG3).

4.4 Legal and Structural Planning Process

The third category of the questionnaire addressed the planning process and legal considerations, asking about the participating stakeholders and often accruing legal and structural challenges.

4.4.1 United Kingdom

Key stakeholders within the redevelopment process were stated the owner, local authorities, preservation officers, and Historic England as a statutory body by all experts (RUK1 et al.). Addressing the owners of the buildings, several interviewees stated the owners to be bodies of the government (RUK2, RUK3, RUK4), developers (RUK2, RUK3, RUK4), and privately owned within companies and family estate (RUK1, RUK3, RUK5). One interviewee stated all industrial heritage refurbishments undertaken in relation to mill projects, mainly situated in the northern part of England, were privately or family-owned, and thus being preserved more sensitively and personally engaged (RUK5). Three experts mentioned that in addition to *LBC* furthermore planning permission is required. Adding challenges arise due to occasional tension between the authorities within the same government bodies or between project developers and government bodies (RUK2, RUK4, RUK5). Two experts stated that good consultants are a basic requirement for projects of historical industrial buildings. Thereby excellent architects, and engineers within mechanic, electric, damp, and building surveyors are necessary to handle the challenges (RUK1, RUK2). While other interviewees who were precisely these consultants, mentioned to be essential within the process. Furthermore, highway consultants (RUK2, RUK5), ecologists, landscape, structural, and drainage consultants were stated as stakeholders (RUK5). Two experts mentioned engaging agencies to achieve a decent mixture of tenants while benefitting from their domestic knowledge and understanding of market demands and rental values (RUK1, RUK5). Additional interest groups and societies mentioned were the twentieth century society (RUK2, RUK3), and the Victorian Society

(RUK3). Regarding the most common legal difficulties, all experts mentioned the *LBC* and *Planning Consent*, which needs to be obtained, adding time expenditure, complexity, and planning effort with authorities (RUK1 et al.). Two experts highlighted the complexity of a change of use of a building, for instance, from industrial to residential (RUK1, RUK2). Four interviewees mentioned occasionally required further surveys to get *LBC* (RUK1, RUK2, RUK3, RUK5). One expert stated severe difficulties as soon as certification of materials becomes demanded by refurbishments. Another difficulty arises by determining responsibility for sections of a structure that had not been restored while other parts have been due to part-owned buildings (RUK3). All interviewees mentioned contamination as an issue dealing with soil cleaning due to former industrial use or contaminated building parts, like asbestos (RUK1 et al.). Two experts highlighted structural challenges regarding dilapidated buildings (RUK4, RUK5). Furthermore, all interviewees mentioned structural and technical challenges with sustainable heating due to an adequate installation of heat pumps (RUK1 et al.). One interviewee mentioned difficulties resulting from historical buildings' natural ventilation being prevented by insulating them (RUK4). Additionally, one interviewee stated the intense and challenging interaction with the authorities lasted five years to achieve a master plan for the whole industrial area development. He highlighted discussing which buildings getting approved for demolishment and which should be preserved. Nevertheless, generating a masterplan with the consent of a certain amount of office, retail, food and beverage, and residential use which the developer can allocate the different buildings themselves as soon as potential tenants enquired rental space, with a concept for around 150,000 square feet of development sites, was perceived as a reasonable strategy with own planning freedom since the masterplan was achieved (RUK1). One interviewee stated that according to residential use integrated into industrial areas, frequent legal conflicts arise. Thereby, residents perceived industrial uses in the neighbourhood due to noise and other nuisances as disturbing and initiated legal action (RUK2).

4.4.2 Germany

All experts pointed out that early involvement, open communication, and cooperation with the heritage authorities are crucial for the successful redevelopment as major stakeholders (RG1 et al.). Two experts also underlined the importance of including politics in order to carry out a successful planning process for large-scale projects (RG3, RG4). Regarding owners, three experts mentioned funds and fund constructions frequently involved in the redevelopment of historical industrial sites (RG1, RG3, RG4). One interviewee especially stated fund management in the *UK* is a frequent investor (RG3). Besides funds one interviewee stated family offices as investors (RG1), while another interviewee mentioned very wealthy private individuals in individual cases, differentiating that smaller residential properties or terraced houses in industrial areas may also arouse interest among wealthy private investors (RG4). One interviewee mentioned municipals as owners of a significant share (RG3). Other stated stakeholders were project developers, architects (RG1 et al.), property evaluators, land surveyors (RG2), structural engineers (RG2, RG4), species conservationists, energy consultants (RG1, RG4), and colour assessors (RG4). Regarding structural challenges, three interviewees mentioned challenges according to fire protection and rescue plans (RG2, RG3, RG4). Furthermore, sound insulation, impact sound (RG2), and challenges in relation to sustainable heating systems such as heat pumps combined with the appropriate locations were stated (RG1 et al.). Other issues mentioned were rust (RG4) light-flooding (RG2, RG4), and ventilation of the old building fabric (RG1, RG3). One interviewee stated according to small-scale regulations, for instance, windowsills, exact window types such as lattice windows, window bars, and even buttons that needed to be specifically used in the interior. The expert stated that these restrictions cause enormous time requirements, coordination, and subsequently rising costs which even might prevent project developments (RG1). Two experts mentioned requirements for colour concepts and necessary colour reports, which also led to enormous time and cost expenditures (RG2, RG4). One interviewee reported a case of discussing a colour concept with the authorities for a long time, subsequently implementing the specified, claimed colour concept and, following the window installation, changing the colour to the applied and planned one at their own expense after consulting with the authorities once more due to poor visual appearance which the authority admitted (RG4). Furthermore, an expert pointed out that although buildings must be ecologically refurbished, in applications, facades may not be permitted to be altered since heritage protection overrules sustainable measures, which leads to controversial discussions (RG1). One interviewer also mentioned two notable examples. In one case, the entire project development was prevented after the heritage authority insisted that the door frames in the building could not be widened by four inches, which would have been necessary for the new use and resulted in the buildings and site remaining unused. In contrast, she mentioned a project that transformed a former industrial woollen site into a park for research energy projects including *PV* which caused several structural changes (RG1).

Three experts mentioned that, in addition to the purely built heritage-related problems, there is often also the hurdle between the authorities, where the building authority involved has different views and requirements, then the heritage protection authority, which contradict each other and in the end the project developers have to act as mediators after they communicate and cooperate inadequately with each other (RG1, RG3, RG4).

4.5 Financial Claims

Section four of the questionnaire asked the interviewees about supporting financial resources and if they feature incentives for redevelopment projects.

4.5.1 United Kingdom

Considering the financial and economic realisation of the redevelopment projects of historical industrial sites all interviewees answered that redevelopment projects rely on self-funding to be able to realise these projects and that very few to rare financial support is granted (RUK1 et al.). One expert stated due to being dependent on self-funding for the development and having no meaningful financial claims, it might probably last fifty years to develop the whole former industrial area. Accordingly, for the first ten years, the development needed to generate revenue streams with commercial use to be able to start the redevelopment and will continue to be a project for an entire generation (RUK1). Another expert made a comparable statement, in which the investment sum of the exemplary overall project is around five to six billion pounds, presupposing wide financial shoulders and resulting in a step-by-step development of the urban area (RUK5). Three experts mentioned financial support might be granted by Historic England with the statements on the scope for buildings at risk receiving support (RUK1), generally rather rare (RUK2), and being under strict conditions for different sectors and little available nowadays (RUK5). Three experts mentioned the National Lottery Heritage Fund granting financial support (RUK3, RUK4, RUK5) with restrictions on third-sector organisations such as charitable organisations, trusts, and non-profit organisations (RUK4), depending on the case and the corporations requesting (RUK5). One interviewee mentioned the *Sect. 106 Agreement* in the *UK* which is about mitigating the impact of the development. Thereby a project developer normally might have to make contributions, for instance, to highway improvements, which can be remitted after the added value is created in the costly heritage refurbishment and the additional costs would prevent the project (RUK2). Furthermore, one interviewee mentioned tax reliefs of the business tax on empty space which are granted for listed buildings, compared to unlisted sites that only receive this benefit for three to six months (RUK1). One expert stated that refurbishment projects are under 20% *VAT*, while the development of new buildings is not, stating it as dis-incentivised (RUK3). One expert mentioned a programme titled *Homes England* which had a fund available for developing and converting heritage sites into residential use (RUK5). One expert highlighted *Enabling Development* for heritage assets, which is non-monetary support but enables the developer to begin to develop specific sections with permission of greater alteration, to generate initial revenue to carry out further development subsequent (RUK2).

4.5.2 Germany

The German interviewees mentioned that there are various and different funding opportunities. One expert listed the levels of the *EU*, the federal government, federal states, and municipal grants for the preservation and energy-efficient refurbishment of constructions and special programmes for built heritage (RG1). One expert mentioned the various possible funding instruments such as grants, reduced interest rate loans, and special depreciation allowances (RG2). The topic of special depreciation was mentioned by all interviewees (RG1 et al.) with the comments made of up to 30% remission of reimbursement (RG1) and a restriction predominantly regarding private investors who are allowed to amortise 10% each year over a period of ten years, with the restriction of not being available for institutional investors (RG4). The interviewees mentioned that these models are no investment incentive for investors and that the incentives for the investment in historical industrial buildings are mostly driven by the location of the buildings and sites with their unique character resulting in economically viable financial investments (RG1, RG3, RG4). Besides the restricted scope, two interviewees emphasised the problem of the availability of funding and that there is no reliability in this regard (RG2, RG3). One of the two mentioned that means of conveyance are frequently depleted until the project development has been planned and building permission is granted, because final planning has to be sent in for an application (RG3). Another expert referred to a project whose funding was dependent on subsidies. However, these subsidies were changed several times, which meant that the project could not be finalised for a long time and had to be constantly rescheduled (RG2). In this matter, the expert highlighted that the projects must be financially viable on their own terms and that the funding can at most be an economic incentive to absorb additional arising costs within the redevelopment of historical industrial buildings (RG2). All interviewees stated that despite various possible instruments of financial subsidies, the efforts for applying and thus the financial added value are not in a reasonable proportion (RG1 et al.). One Expert stated that this could only be of interest to private households who are already very wealthy anyway (RG4).

4.6 Redevelopment and Its Challenges

The fifth section of the questionnaire concentrated on the redevelopment and the arising issues related to the construction and site. Additionally, the possibility of a redevelopment project that is economically viable and the extent of cooperation with the authorities were both questioned.

4.6.1 United Kingdom

Four interviewees mentioned vandalism and natural deterioration as frequent issues and leading to poor conditions of the buildings as a challenge (RUK1 et al.). One interviewee provided examples of low-level antisocial behaviour that occurs in abandoned areas, such as the destruction of windows and arson (RUK1). Another expert gave an example of decades of vandalism with arson and cars being burned at a site while mentioning the mitigation of vandalism throughout the refurbishment (RUK5). In contrast, one expert stated that historical industrial sites are more often premium situated and that it is no bigger deal than anywhere else (RUK3). Besides that, all interviewees highlighted contamination as a significant issue often appearing. Additionally, all experts mentioned soil problems (RUK1 et al.) and two of them stated problems appearing with mills that had dye houses and pits on their sites (RUK1, RUK5). Thereby one expert mentioned a company whose business model was to decontaminate sites because they discovered the absence of developers interested in these sites and at the same time clearance not being as expensive as everybody expected (RUK4). Naming further issues, one expert mentioned mine shafts and wells shafts frequently emerging under buildings in areas of coal mining and representing bigger problems than soil works. Additionally highlighting those different challenges and knowledge is necessary in areas where coal, woollen, rubber, or cargo has been predominant (RUK5). Besides the contamination of the ground, all interviewees mentioned built fabric contamination such as asbestos as very common which leads to repeated problems, extra work, and costs (RUK1 et al.). One expert thereby mentioned plenty of discussions with authorities due to the demand of removing asbestos, but at the same time not being allowed to harm any construction parts (RUK2). Another issue mentioned was protected species and animals living at abandoned sites and in buildings that have been vacant for a long time. Thereby the interviewees mentioned animals like protected birds, bats, newts, owls, and peregrine falcons as frequent problems (RUK1, RUK2, RUK3, RUK4,). One expert mentioned that canal sites frequently cause problems due to the flight paths of birds and construction sites being close to water. Even changing lights often cause immense issues due to animal welfare (RUK2). The unexploded ordinance was stated as a minor problem with the redevelopment of the sites by most experts (RUK1, RUK3, RUK4, RUK5). However, one expert highlighted the enormous issues arising when they appear assigning the issues occur more in the south than the north of the *UK* (RUK2). Two experts mentioned that unexploded ordinance is a topic that is proved within a purchasing process and can be examined in existing databases (RUK3, RUK4) while one expert mentioned already most being eliminated (RUK4). Experts furthermore mentioned that a developer must understand the building's history, the building's fabric, and specific construction parts of the building. If an in-depth comprehension of the construction has been achieved, it is possible to generate concepts that enable an advanced planning process with authorities (RUK1, RUK3, RUK4). All experts highlighted the crucial necessity to cooperate with the authorities at an early stage (RUK1 et al.). One expert highlighted the pre-application procedure that is accessible to create a joint plan and get approval, thus promoting good cooperation (RUK2). One interviewee mentioned that the authorities are supportive and encouraging because of their development plan to redevelop the industrial sites, revive the area, and generate employment in the area (RUK1). Another interviewee had similar experiences and added that authorities contributed to the process and design and had good ideas themselves. Furthermore, they prompted directions, and compared to similar projects. Furthermore, he stated that, cooperating early in the planning process they start less acting as police and start to think in line (RUK4). Other experts highlighted heritage authorities' work helps to investigate and understand the building, and its history and provides valuable information for refurbishment which thus might be useful for commercialisation (RUK3). All experts stated that the decentralised system is promising, in which the local authorities have domestic knowledge and can enforce and implement the developments with centralised general regulations provided (RUK1 et al.). Focussing on economic viability all experts stated that revitalisation, management, and a new use concept are the best way of protection and preservation. Thus, always a case-to-case strategy is required which utilisation concept might be suitable (RUK1, et al.). Two experts highlighted the commercial value proposition depends on the developer and the perspective on *ROI* whether it is short-termism versus long-termism in relation to the developer, investor, or owner-operator (RUK3, RUK5).

4.6.2 Germany

Facing the challenges of external influence, two interviewees stated vandalism as a common issue causing additional costs and harm to the buildings and sites (RG1, RG4). One of them mentioned issues of squatters, drug dealing hubs, and spray painting. However, bigger problems were linked to people who stole scrap metal and copper cables causing harm to the fabric of the building (RG1). Two experts stated vandalism as a less significant concern since, for instance, if detachable components are damaged by spraying, these parts will be replaced or revised during the revitalisation process anyway (RG2, RG3). One interviewee stated that vandalism is visible and assessable like arson and broken windows that frequently need to be replaced anyway (RG2). On the other hand, three experts categorised natural decay and environmental influences as more significant problems (RG1, RG2, RG4). Thereby, problems with long-term vacancy are plants and tree growth attacking the substance and causing cracks and break openings (RG2). One interviewee questioned the degree of general damage when water leaks affect the structure and whether preservation might be feasible (RG4). Furthermore, two interviewees addressed emerging biotopes and the settlement of animals such as birds, hares, newts, and ground breeders as significant problems (RG1, RG2). One stated these issues might even cause development prevention (RG2). Another interviewee stated that ownership of properties entails obligations under the law and that in general damage to third parties and decay must be prevented (RG1). Besides that, a major obstacle highlighted by all experts was the contamination of soil or building fabric, which is difficult to predict, and if occurs, leads to immense costs and time expenditure (RG1, et al.). In this regard, all interviewees mentioned asbestos and individually mentioned biphenyls (RG3) and synthetic mineral fibre (RG4). One specialist additionally addressed the significant risk posed by unexploded ordnance still buried in the ground (RG2). Regarding these risks, all experts rated these problems occur more frequently than for other project developments (RG1 et al.). Concerning challenges and obstacles, three interviewees made extremely critical statements about heritage protection authorities and called for greater involvement in the processes (RG1, RG3, RG4). One interviewee mentioned that authorities are legally not obliged to help and often only accept or reject both building and heritage applications with alterations. Adding that authorities sometimes do not operate in an outcome-orientated manner and rarely incorporate their experience of similar cases. He also mentioned that the heritage authority had even refused other authorities a planning application for their building after a barrier-free lift was not integrated into the external appearance of the building (RG1). Another interviewee also stated heritage authorities act as application centres and depicts some cases as a ping-pong of applications with challenging interaction and communication (RG3). The third critical expert mentioned and criticised inconsistent actions of the authorities and emphasised the different opinions between upper and lower heritage protection authorities within the regulatory system (RG4). In contrast, one expert mentioned heritage authorities acting inconsistently due to the federal system but he has experienced both cooperative and outcome-orientated authorities as well as authorities where many compromises had to be found (RG2). In order to generate an economically viable redevelopment of historical industrial sites, all experts emphasised intense interaction with authorities and the creation of a reliable, reasonable, and sustainable concept with mixed-use integrated within urban development as a determinant of success (RG1 et al.). Additionally, one interviewee stated the integration of neighbours and politics is necessary (RG4).

4.7 Conclusion and Trends

The interviewees were asked to provide information on the opportunities and trends associated with the revitalisation of historical industrial sites, as well as whether it would be a feasible concept for *ESG* developments, in the final category of the questionnaire. Additional traits and concepts were requested in the final question.

4.7.1 United Kingdom

All experts highlighted the uniqueness, aesthetics, character, and architectural design and associated the redevelopment of industrial historical buildings as pleasing places (RUK1 et al.). Four experts stated these redevelopments often drive community engagement and integrate the regional communities while focusing on the heritage aspects of the historical industrial buildings (RUK1, RUK2,

RUK3, RUK5). One interviewee stated this creates a sense of belonging among tenants and visitors within the developed site (RUK1). Those developments were associated with the creation of public spaces and generating urban realm by three interviewees (RUK2, RUK3, RUK5). Additionally, it was stated that place-making generates cultural anchors (RG3), and focal points for a broader regenerative project where there are possibly more lifeless, insignificant modern buildings situated in the neighbourhood (RUK2). A trend associated with redevelopments by three experts is the sustainability and *ESG*-movement (RUK3, RUK4, RUK5). Thereby on the one hand the trend of refurbishing instead of demolishing was mentioned (RUK4) and the so-called *B-Corps* movement, enhancing the people, the planet, and profit at the same time (RUK3). Moreover, the trend of regenerative architecture (RUK2, RUK3) and regenerative design in relation to bird bricks, bee bricks, and biophilic design was mentioned (RUK3). Furthermore, the term adaptive reuse was used by two interviewees according to the design and architecture merged with sustainability (RUK3, RUK4). Two experts highlighted that rebuilt sites with their historical backgrounds are more valuable and connected with well-being, mental health, and are regarded as pleasant working environments, especially after the *COVID*-pandemic. (RUK1, RUK5). According to the statements they meet social responsibility and go alongside co-working concepts (RUK1, RUK3). Furthermore, one expert highlighted that people want to represent their standing and feelings. Thereby sustainability can be transmitted through offices and businesses situated within historical industrial sites that attract people and especially the younger work generation (RUK3). Two interviewees stated the refurbished historical industrial buildings combine commercial and social benefits (RUK1, RUK2). The phrase 'creative placemaking' used by one expert combines elements of heritage, culture, and commerce. Thereby the interviewee added to create valuable income through the cultural part with, for instance, hosting events which is advantageous for the historical industrial buildings and would not be possible with a 'modern tin' development (RUK1). Three interviewees mentioned a more creative use and tenant mix with crafting and artistic tenants such as gin making, breweries, wood crafting, jewellery, weaving, art, and art galleries (RUK1, RUK2, RUK5), linked to the bohemian trend (RUK2). The growing trend of educational use and universities as tenants was stated by one expert (RUK5). Furthermore, one interviewee stated the flexibility of these buildings and used figuratively the term 'cart-horse' for them. This implies the simple adaptability and changeability of the buildings to new uses as unique characteristics. Highlighting these constructions have already existed for decades, have been converted several times, and are capable of further conversion based on reliable building structures (RUK3). Experts mentioned that cultural elements and mixed uses attract people (RUK1, RUK2, RUK3). With this attraction further revenue streams can be generated with events such as Christmas markets, fairs or even serving as a filming location stated one interviewee (RUK1). He mentioned a bottom-up creation of the developed site is possible with social communities and associations being attracted and approaching the developed sites both as event locations and interested tenants (RUK1). Thereby one expert mentioned the social value gaining immense relevance due to the existing *Social Value Act* and furthermore the upcoming *Procurement Bill* in autumn 2024 (RUK4).

Approaching a more philosophical view one expert has posed the "parable of Theseus ship [...] from Greek mythology. It is this idea that if you replace a part of your ship year on year over a period of 20 years, so much so that you [...] have ended up replacing every element of touch. Is it the same ship?" (Interviewee, C.1.3, minute 22:34). Thereby he questioned the conservation and highlighted, that older generations have a different perspective on what we call historical industrial buildings today. For some, thesem buildungs were places of employment, whereas for younger generations they are increasingly just reused factories (RUK4). Furthermore, he stated that in Western Europe we act with an authenticity to the materials used, compared to the Middle East which does not care about the exact material but on the workmanship and hence reconstructs buildings in the same way, but with different materials. He thereby questions whether the historic value is in the fabric itself or in the architectural appeal of a heritage building (RUK4).

4.7.2 Germany

Facing trends for the redevelopment of historical industrial buildings, all experts emphasised the relevance to urban development and neighbourhood development (RG1 et al.). Within urban development, one expert mentioned the aspect relating to *AI*-supported urban planning systems of the urban realm and movement patterns through automation systems of buildings and management systems (RG1). Two experts highlighted the relevance of urban development according to new mobility concepts that can be considered and integrated in the creation of new urban areas (RG1, RG4). In conjunction with the urban realm, two interviewees emphasised the topic of human protection and environmental influences, such as heavy rainfall and the integration of large areas as retention spaces with installed cisterns. Furthermore, the prevention of heat islands while establishing cool places with green spaces within the cities was mentioned (RG1, RG2). In the matter of urban development, the aspect of re-densification of the urban realm was stated by all experts as a significant opportunity in relation to already overbuilt areas without new land consumption (RG1 et al.). In this regard, three experts also mentioned that due to the development of the market with increased construction costs, the refurbishment of buildings is more favourable than the demolition and construction of new bare brickwork (RG2, RG3, RG4). In addition, all interviewees mentioned the subject area sustainability as a high priority within the redevelopment of former industrially used historic buildings (RG1 et al.). However, one expert mentioned in the context of sustainability that preservation is not the most sustainable strategy in all cases, depending on existing structural conditions and contamination. Furthermore, adding the recycling of building blocks often involves considerable expense, which occasionally generates a comparable carbon footprint (RG4). Thereby the carbon footprint was also mentioned by one expert with reference to the fact that the real estate industry causes 40% of carbon emissions. In addition, the interviewee mentioned *CSRD* statutory reporting requirements by prioritising sustainability and the overall ecological footprint. Highlighting that companies that do not achieve the certificates will be incapable of action due to regulations (RG1). Apart from that, one interviewee emphasised the structural change of the retail industry and a shift of emphasis towards liveability of new retail concepts. Consequently, buildings that create liveability generate particular attractiveness for the retail, commercial, and residential sectors. He stated that this creates increased pedestrian frequencies in the retail area, which hence can increase sales and rents. The expert furthermore mentioned that increased pedestrian frequencies caused by such redeveloped historical buildings might be able to compensate for the loss of rental area due to unfavourable floor plans with extensive areas for common use, wide stairwells, or unusable floor space with attractiveness for pedestrians and visitors (RG2). Two experts mentioned the special static structures of the former industrial sites and emphasised the possibility of integrating various use types (RG3, RG4). One expert stated that a suitable mix of uses can promise success through the flexibility of the constructions and long-term adaptability if the buildings and the site are perceived as an overall concept (RG4). In Addition to sites in prime locations driven by investors with a monetary calculation, one interviewee mentioned that peripheral locations might also profit from the current trends. According to the expert, other investment ideas might result from a more personal connection to the substance as well as an emotional historical connection. These are often museum uses merged with a stronger local economy and social responsibility from previous or contemporary business contexts (RG3).

5 Discussion of the Conducted Research

5.1 Expectations

To investigate different approaches to handling and developing historical industrial buildings between Germany and the United Kingdom, qualitative research using semi-structured interviews based on a questionnaire with six categories was conducted. To examine possible differences, experts in both countries who are involved in the development of historic industrial buildings were interviewed to obtain reliable statements with insights into professional practice for both nations.

The research was anticipated to identify different approaches in handling and potential development strategies for historical industrial buildings, if available, and to derivate methods or approaches for the respective other country for further project developments of the special asset class.

The analysed special asset class of historical industrial buildings presents a variety of heterogeneous challenges as well as some opportunities. The relevance of the undertaken research arises from the transformational process of cities being affected by, for instance, re-urbanisation and new demands for urban development such as urban resilience and sustainability (See 2.5, 2.6). Merging the challenges with the fundamentally similar and related history, initiated by the *Industrial Revolution* in the *UK* (See Chapt. 2.1), evaluating the management of historic protected buildings may provide valuable insights into the potential replication of strategies by the other nation in the contemporary era, as was done centuries ago.

In order to fulfil the quality standards of qualitative research, the scope could be met by the participation of five qualified experts in the United Kingdom and four equivalent interviewees in Germany, each of whom had a distinctive

R. Groer et al., *Historical Industrial Buildings and their Real Estate Utilisation*, Studien zum nachhaltigen Bauen und Wirtschaften,
https://doi.org/10.1007/978-3-031-89417-6_5

viewpoint on the redevelopment process in their respective countries. The interviewee's quality was met according to their average professional experience of twenty-four years. The expert with the least professional experience had fourteen years of experience, while the expert with the most professional experience had forty years. Experts with extensive professional experience, not restricted to a single region, submarket, or professional field in both countries, were selected to gather information and perspectives on the development of historical industrial buildings from a diverse range of experiences (See 4.1). This was carried out to guarantee the reliability of the statements achieved. In addition, the procedure within the thesis was described in detail to ensure transparency of the research, and the views of the respective interviewees are presented and stated extensively and comprehensively within the presentation of the results.

5.2 Discussion and Evaluation

5.2.1 Historical Background

Conducting a comparison between the *UK* and Germany based on the differences regarding the historic industrial buildings, the first distinction can be identified in the buildings' and sites' legacy. The *UK* initiated the *Industrial Revolution* in 1760 (Stearns, 2021, pp. 42–7) and started to spread out the revolution which arrived in Germany with the importation of the first machinery around 1850 (Allen, 2009, p. 2). Besides initiating the whole movement, the *UK* invented the iron-framed construction technology and accordingly, the first iron-framed mill, named Shrewsbury Flaxmill, was constructed and still persists (Clarkson, 1990, p. 50). With the invention of the first automated spinning machinery, the *UK* established itself as the global leader of the woollen industry which led to the fact of numerous diverse mill buildings existing. Germany, on the other hand, was initially known for the hard coal and iron industry, the Ruhr being a famous example. Later on, it was recognised in the sectors of mechanical engineering, electrical, and chemical industries. (Hahn, 2011, p. 27). Thereby a distinction between the two countries arises from the age of some still existing constructions as well as the variance of building types, linked to the industries being predominant. Nevertheless, mills are a buildings that were frequently mentioned in both nations, with a variety of mills existing in Germany, such as flour mills, hammer mills (Wild, Denkmalschutz Kompendium, 2023, pp. 82–83), and even more often stated in the *UK* with factory mills of woollen and cloth industry (Allen, 2009, pp. 183–184). RUK5 also mentioned the variety of mills and added that all of the converted mill projects undertaken, that were mainly located in the North of England, were still in private or family ownership. Adding the sites are frequently located not only in the major cities but also in more rural areas and have been converted with particular care and consideration by the owners with a high level of personal involvement (RUK5). In the final outlook question, a German expert stated that industrial sites in economically underdeveloped areas, still owned by the former enterprises' families might benefit from the families' involvement in undertaking developments. He mentioned that these developments are executed with a different personal interest than those of professional investors focusing on financial calculation. This may result in an important factor for the type of buildings within more community-based rural areas (RG3). One of the interviewees from the *UK* is a project developer, as described by RG3, and stated the strong personal relationship to the buildings, which are still family-owned. Hence, he and his business partner are trying to only undertake the most necessary interventions in order to preserve the existing buildings and the family legacy (RUK1). He mentioned the approach of 'creative placemaking' using elements of culture, heritage, and commerce while creating local jobs after the factory was closed (RUK1). The approach of development and the utilisation concept strongly involves and emphasises the incorporation of a regional social environment as tenants and thus creates a local community. RUK1 stated to attract people with the development, which is adding additional value with art programmes, handcrafting, various events, movies being shot, and for instance, the social running clubs weekly coming around stimulating food and beverage tenants (RUK1). Thereby RUK5 mentioned similar projects with mills and the involvement of the local communities and furthermore stated educational facilities as potential tenants (RUK5). Combining the statement of RG3 in the German market, in which he sees potential in buildings in peripheral locations and economically underdeveloped areas as development opportunities with the higher personal engagement of locals with the concepts described by RUK1 and the confirmation and additions by RUK5, a utilisation concept for project developments in less economically prosperous areas might be derived.

5.2.2 Heritage Protection

Comparing the heritage system, Germany having a federal state system, and the *UK* consisting of four nations, there are similarities in the specification of regulations passed on from hierarchical levels from the highest body of the governments to the local authorities (Fig. 2.2 & Fig. 2.3). At this juncture the listing of heritage in the *UK* is based on a national criterion within the four countries, each with their own list including all built heritage. Regarding England, there is *Historic England*, a government body qualified to manage the listing process (Taylor & Sorensen, The Building Conservation Directory 2021, 2021, p. 25). In comparison, in Germany, this responsibility is assigned to the respective federal states and there is no independent authority that carries out a categorisation or listing for the entire country. Facing the individual protection of a building, there are noteworthy differences. In the *UK* there is a listing system using grades for a distinction between different protected buildings (Taylor & Sorensen, The Building Conservation Directory 2021, 2021, p. 25; Tbl. 2.4; RUK1, et al.). A distinction is made between twenty categories with two hundred subcategories, provided by *Historic England* in a publicly accessible document (Historic England, Scheduled Monuments, no date). Comparing it to Germany there is only one type of protection, no grade system, and hence no such document. Regarding the scope of protection, both countries got an area-based protection which is the conservation area for the *UK* and the heritage ensembles in Germany. According to the literature, in the *UK* the whole building or parts of the building, including interior and exterior, unless they are explicitly excluded, are protected (Morrice, 2021, p. 13). While in Germany, according to literature and expert statements, various distinctions can be made between landscapes, the ensemble, entire buildings, parts of buildings, halls, and interiors (Wild, Denkmalschutz Kompendium, 2023, pp. 129–134). The interviewees even stated buttons of the interior (RG1) and colours are considered (RG2, RG4). Thus, it can be derived that there are similarities in the area-wide, and exterior protection, but differences when it comes to single buildings and some mall-scale heritage approaches by Germany.

Besides that, the *UK* has a second listing focusing on sites at risk of decay due to abandonment, known as *Heritage at Risk* in England (Historic England, What is the Heritage at Risk Programme?, no date). Furthermore, local authorities are in charge of *LBC* in the English heritage system, and *Historic England* acts as an advisory body for authorities and developers (Historic England, Listed Buildings, no date). In contrast, Germany has a federal structure and federal states with varying laws and regulations for the protection of built heritage. Hence the federal states with local authorities are in charge of listing buildings with thereby slightly varying methods of classification for built heritage. In addition, the German local authorities are responsible for the change applications for built heritage, which must be submitted and often require additional planning applications (Wild, Denkmalschutz, 2024, pp. 28–29). Facing the structure of heritage protection, the fundamental governmental structures are comparable, but the implementation varies. Using a grading system with a specifically assigned government body distinguishes the *UK* from Germany with each federal state being in charge. Another distinction is that the *UK*, unlike Germany, has a second listing of buildings under risk created by the government body. A similarity between the two countries is the special authorisation which is demanded by local authorities when alterations and adaptions are planned. In both countries, in addition to special heritage planning permission and the *LBC*, general planning permission by local authorities might be required (Morrice, 2021, p. 11) (Wild, Denkmalschutz, 2024, pp. 28–29). In Germany and the *UK*, the collaboration between heritage authorities and developers, who are both primary stakeholders of developments, is an area of varying perspectives. Evaluating these statements, it must be taken into account that there are local authorities with different responsible heritage officers in different areas and nine reports on the interviewees' experiences with different authorities. The professionals in the *UK* stated the cooperation as additional work. One expert stated the pre-application process is a useful tool to sound out the planning in cooperation and develop refurbishment together (RUK2). The interviewees from the *UK* stated that local authorities frequently exhibit a lack of resources for more intensive support and that their decisions are often subjective and inconsistent. Nevertheless, the interviewees stated that cooperation with the local authorities is often cooperative, authorities are sometimes participating in the planning process and provide historically useful documents, if they are involved at an early stage and the consequences of adaptions are considered carefully (RUK1, et al.). Regarding cooperation with the authorities in Germany, as noted from the statements of the interviewees, three out of four stated negative previous experiences, or at least mentioned them more frequently and examined them more intensively than cases that were less challenging (RG1, RG3, RG4). All interviewees mentioned the subjective views of heritage officers (RG1, et al.) and one mentioned that, unfortunately, the heritage officers are legally not obliged to consult and therefore often only reject or approve applications (RUK1). Thus, it might be inferred that cooperation in the *UK* with the heritage protection authorities seems to be more cooperative and target-oriented. However, it should be borne in mind that the German interviewees who were critical also stated successful project developments, if there was very early and intensive co-operation with the authorities. It can be concluded from the statements that the involvement of the authorities in the planning process from the outset is crucial to the success of a development project. Without wishing to generalise these comments to the whole market, it might be concluded that cooperation in the *UK* seems to work better than in Germany.

Asking the interviewees if there were any facilitated processes for developments, on both sides all experts answered no. Considering the different types of listing and linking them to statements given, there are approaches that could be perceived as simplifications in an international comparison. The *UK* distinguishes built heritage using grades and according to the grading system, no alterations other than preservation in *Grade I* up to permitted alterations combined with extensions and adaptions from *Grade II** and especially in *Grade II* are authorised, as stated by various experts (RUK1, RUK2, RUK4, RUK5) and even by the interviewee who is conservation accredited (RUK3). Highlighting the stated ensemble protection by RG3 which maintains external appearance and allows new construction attached, which was not mentioned by *UK* experts or reviewed literature, but the concept of 'facadism' (Bargery, 2005) was found to be debated on the basis of subsequent research, it might be perceived as the existing simplified process by the *UK*. According to those examples and differentiated methods of protection, both countries could consider the concepts as possible approaches. However, the introduction of a grading system could raise the question of prioritisation and at the same time the problem of discrimination between listed buildings. As it was described in the debate of RUK3 after the idea of introducing *Grade III* as protection for all existing buildings being discussed due to reasons of sustainability.

5.2.3 Financial Claims

The funding of project developments is based on the creation of an economic concept with a mixture of uses, as stated by all experts, both in the *UK* and Germany (RUK1, et al.). The *UK* interviewees stated that there are no meaningful financial claims available and if, they are uncommon and restricted (RUK1, et al.). Although financial incentives for the inheritance of listed buildings were mentioned in the literature (HM Revenue & Customs—GOV.UK, 2022, p. Chapter 2), neither RUK1, who is running a family-owned company, or RUK5, who mentioned various cases of mills projects in private or company ownership, mentioned existing relevant financial incentives. Nevertheless, it was mentioned, that possible meaningful financial grants might only be granted by the *National Lottery Heritage Fund* (RUK3, RUK5) which was also stated to be restricted to mostly charitable organisations (RUK5). One incentive for listed buildings mentioned was the exemption from property tax if they are untenanted (RUK1). One expert furthermore mentioned a rather rare funding for a project of 5.7 million pounds by the *Levelling Up* fund for a heritage asset by the government (RUK5). In contrast, one interviewee stated refurbishment projects are under 20% *VAT* while new developments are not (RUK3). Derived from the German literature report, there are various financial claims such as special depreciation for meaningful preservation work, state funds, municipal funding, subsidised loans by *KfW* for efficiency house heritage buildings, and subsidised loans (Wild, Denkmalschutz Kompendium, 2023, p. 77). All German experts rated the impact of financial grants as relatively low and insignificant compared to overall costs (RG1, et al.). Furthermore, strict regulations with a limited scope of application exist and grants not being reliable for calculations (RG2, RG3).

Comparing the financial claims, there are several possible financial grants available which all together in each country do not seem to be an incentive to undertake developments. Although none of the German interviewees mentioned and it was not mentioned in the literature reviewed, after RUK1 stated property tax exemption, further research was carried out by literature and talking to a German tax consultant, it was possible to figure out that a reduction in property tax for listed buildings between 10–25% is possible, varying between federal states (Grundsteuer, no date). However, interesting practices to be derived from the above examples for Germany would be the exemption of property tax on untenanted listed buildings and buildings in general for up to three months, as well as the extension of subsidies such as those provided by the *National Heritage Lottery Fund*. Unlike the taxation of renovations and the support of new construction, German financial claims such as low-interest loans and subsidies for sustainable conversion initiatives could be possible approaches for action in the *UK*.

5.2.4 Challenges of Development

Considering the structural difficulties in connection to redevelopment, in general, comparable technical challenges were mentioned which can be related to similar technical developments and construction techniques in both countries at that time. Both the *UK* and German interviewees emphasised that a thorough understanding of the buildings is crucial for success redevelopment projects. Thereby buildings in both countries were found to have structural issues, including soil contamination and the presence of asbestos (RUK1, et al.; RG1, et al.). The German interviewees mentioned additional specific concerns regarding materials such as biphenyls (RG3), synthetic mineral fibre (RG4), and issues related to project prevention, such as biotopes (RG2) and the prohibition of adoptions, such as door frame widening (RG1). Whereas the *UK* interviewees mentioned surplus special issues, such as mine and well shafts (RUK5), developments at canal sites (RUK2), and dye houses at mills (RUK1, RUK5). Besides that vandalism was identified as a common problem, whereas two German interviewees disregarded its significance, stating that it only causes damage to the surface, which needs to be edited in most cases anyway (RG2, RG3). An interviewee from the *UK* described it as a crime that is not considered serious but is still annoying (RUK1). Additionally, the interviewees on each side reported numerous industrial sites that were confronted with the challenges of natural decay and environmental influences, which were identified as a more significant issue (RUK1, et al.; RG1, RG2, RG4). Nevertheless, all interviewees emphasised the buildings' reasonable load-baring capacity and robust static constructions for redevelopments as advantages (RG1, et al., RUK1, et. al.). However, in some specific cases the structures cause problems due to dilapidated buildings (RUK4, RUK5, RG3, RG4), challenging floorplans (RUK1, RUK2, RG2, RG4), leaking, and disadvantageous north-light roofs (RUK1, RUK5). The interviewees stated that case-by-case solutions must be developed in order to address all general and uncommon issues. Thereby the German experts pointed out a greater number of project-preventing challenges, such as protected animals (RG1, RG2), construction alterations (RG3), or investors avoiding regions due to critical authorities (RG4), in comparison to the *UK* interviewees, who mentioned challenges that appeared to be solvable. The concepts of mixed-use development that include office, retail, and residential elements were highlighted by all German and *UK* experts (RUK1 et al., RG1 et al.). Only a few stated light and less industrial uses after a redevelopment as returning strategies (RUK2, RUK5, RG1). In contrast to this, some *UK* experts stated creative concepts and community engagement as successful redevelopments (RUK1, RUK2, RUK5), while furthermore, two *UK* experts mentioned educational uses as promising (RUK4, RUK5). All German experts mentioned project developments with the utilisation mixture of office, residential and retail use. In contrast, the *UK*'s strategies were supplemented with innovative concepts of craftsmanship, artistic concepts, and usage that were extended with community-based strategies, and educational institutions. Thereby these additional mentioned *UK* concepts could be potential approaches for redevelopment strategies for Germany.

5.2.5 Vision and Trends

Several experts repeatedly mentioned the most effective method of preserving heritage buildings is to manage and utilise them, which is in line with maintenance and refurbishment. They furthermore stated that adaptions and repurposing of the buildings and sites often manage to emphasis or even reinforce the importance and value of the construction (RG1, RG4, RUK2, RUK3, RUK5). Thereby one German interviewee highlighted the *Uffizi Gallery* in Florence, which initially served as the office of the Florentine administration and became one of the world's most famous museums for art because of the adjustment to changed requirements and accordingly through a change of use. He stated that it is unfeasible to freeze certain buildings while cities at the same time are in an permanent transformational process. Adding that while it is significant to ensure the safeguarding of some buildings, it is also crucial to ensure new uses that are currently demanded (RG4). The statement about the *Uffizi Gallery* can be adopted in London regarding the redevelopment of the former industrial power station with the extension of the building, which subsequently became internationally recognised as the *Tate Gallery of Modern Art*. Regarding extensions, an English expert referenced one of Germany's most significant building, the *Reichstag*, which was not subject to industrial redevelopment but nonetheless extended with a dome despite its extraordinary national significance (RUK3). To emphasise the importance followed to their redevelopment and to serve for discussions for possible adaptations, both experts mentioned these projects as reference objects. With that, experts on both sides showed the relevance of future developments as potential.

Regarding historic industrial buildings, interviewees both from Germany and the *UK* emphasised the uniqueness, aesthetic, character, and architectural design of the buildings and with that the potential for redevelopment (RUK1, et al., RG1, et al.). All interviewees additionally highlighted mostly attractive locations of these buildings as an attractive factor (RG1 et al.). Combining these two qualities, all German experts stated enormous potential in the urban and neighbourhood development of historical industrial sites. The developments were designated as crystallisation points (RG4) and as a general benefit for the citys (RG1, RG3). The interviewees from the *UK* likewise mentioned the potential of urban development and the urban realm and characterised them as cultural anchors (RUK3) and focal points compared to other modern developments (RUK2). Accordingly, both nations assign enormous potential for urban development by refurbishing the concerned buildings. In contrast to the English experts, German experts furthermore added the relevance of integrable new mobility concepts (RG1, RG4) and developing resilient cities during environmental influences such as heat waves and heavy rainfall (RG1, RG2). Besides that, the interviewees on both sides emphasised the sustainability factor by mentioning regulatory requirements. One German interviewee referred to the *CSRD* statutory reporting requirements, which prioritise sustainable practices, and the real estate industry, which is accountable for 40% of carbon emissions (RG1). One *UK* interviewee featured the

B-Corporation movement, which is committed to improving the quality of life for people, the planet, and profit (RUK3). Furthermore, the *Social Value Act* and the upcoming *Procurement Bill*, which is scheduled for publication in the autumn, were addressed (RUK4). About increased technological sustainability, the *UK* interviewees furthermore mentioned regenerative architecture (RUK2, RUK3) and regenerative design, such as biophilic design and bird bricks (RUK3). Additionally, the significance of adaptive reuse (RUK3, RUK4) and renewable energy and sustainable heating systems were emphasised (RUK1, RUK2, RUK3, RUK5). Thereby new technologies with environmental heat pumps suitable for built heritage were mentioned on both the German and *UK* sides (RUK3, RG2). Likewise, all German interviewees identified sustainable heating and renewable energy as driving factors addressing sustainability (RG1, et al.). On top of that, German experts emphasised elements such as the densification of the urban realm, the prevention of demolishment through the reuse of buildings, and the avoidance of new land overbuilding by developing existing building areas (RG1, et al.). Given that, both countries refer to the enormous trends of sustainability and social impact.

Challenges to the third-party usability was mentioned, especially applying to the unusual properties of historic industrial buildings (Leopoldsberger, Everling, & Salostowitz, 2022, pp. 147–148). Regarding the technical conversion capability, interviewees of both countries stated the buildings' robustly constructed structures as a benefit in terms of technical adaptability (RUK3, RUK5, RG2, RG4). Thereby one *UK* expert stated the load-bearing brickwork walls with load paths are indeterminate and therefore allow structural changes. Additionally, he noted that these structures have rather unremarkable facades. He characterised the historical industrial buildings as adaptable 'cart-horses' that have undergone numerous changes of use (RUK3). Adding that high ceiling heights, and reasonable load-bearing capacity provide them both attractive and suitable for further transformations to various uses (RUK3, RG4). After the building structures are even perceived as advantageous in certain instances, the third-party-usability of the buildings is more or less only dependent on the utilisation concepts. Thereby all stated mixed-use concepts of residential, commercial, and office as integrable (RUK1, et. al; RG1, et al.). While interviewees emphasised the attraction and charm of the developed historical industrial buildings and one German expert stated the retail sector is undergoing a structural change with a trend and demand for retail spaces which supply a kind of adventure, mixed with leisure opportunities within the buildings. He stated increased pedestrian frequency that can result from the associated charm, appeal, and attractiveness of redeveloped historical industrial buildings might consequently increase both the achievable rents for developers and the turnover for tenants (RG2). Consequently, the buildings may even result in a boost in economic efficiency. According to an expert in the *UK*, the integrated 'creative placemaking' concept has contributed to an increase in value, as the demand for rental spaces has increased and new revenue streams have been established, despite the buildings being situated in the outskirts. The interviewee emphasised the sites' and buildings' unique character which turns the place into attractiveness. Highlighting, that this effect would not have been able to achieve through a new construction rather than a refurbishment (RUK1). On top of that the 'creative placemaking' approach (RUK1) represents a utilisation concept incorporating a community-based approach, and thus is in line with the sustainability trend and emphasising social impact.

Furthermore, respondents from the *UK* asserted that the combination of conveyed values and appearance causes attraction. As a result, tenants and employees of related companies are attracted to the buildings due to the representation of their positions which is increasingly emphasised (RUK3). One interviewee stated that this is particularly appealing to the younger generation and that enterprises utilise it for external perception, which can on top be advantageous for recruiting (RUK3). Thus, one *UK* interviewee mentioned the *bohemian* trend, which attracts handcrafting and artistic tenants integrated into rental mixed structures in historical industrial buildings (RUK2). In addition, educational uses, such as universities were also described as a trend for tenants in the *UK* (RUK5).

Possible approaches and a strategy for a utilisation concept might be derived for Germany based on the aforementioned more creative use concepts, the structural change in the retail sector, and the need for retail spaces with adventure and leisure-appealing aspects. This possible mixed-use strategy should incorporate the 'creative placemaking' approach in creating a tenant mix and hence include handcrafting and artistic tenants to incorporate the social aspect. By considering this approach besides the structural changing retail sector (RG2), as well as more outskirt situated buildings and sites, stated as a challenge (RG3), might benefit and thus increase in values to make a project development economically viable. Thereby increased pedestrian frequency and attracting enterprises and their employees might conclude to a reasonable, financially viable concept.

While at the same time considering the social impact, which could be especially beneficial for the challenging planning process with authorities, the issue both with buildings located in the outskirts and family-owned buildings with different development expectations (RG3) and the demand for retail spaces with an experience factor (RG2) could be addressed by using a combination of the uses mentioned above of 'creative placemaking' (RUK1).

5.3 Limitations

The reader should bear in mind that this study is based on expert interviews with answers generated using a virtual communication tool which influences the interactivity and the flow of the conversation. Conducting interviews might tend to present the own performance, company, or native country in a more positive light.

Due to practical constraints, this paper cannot provide a comprehensive review of all the specific examples given within the interviews due to a limited scope of extremely individual projects and thereby the problem arising to effortlessly identify the interviewees. Another potential problem is that the scope of the thesis may be too broad due to the variety of comprehensive topics, starting with individual historical industrial buildings some of which date back to the *Industrial Revolution*, the area of heritage protection, and the subject area of project development and revitalisation in two countries.

Furthermore, the limited time frame for the research in the context of an international comparison limited the depth and scope of the research by identifying and scheduling appointments with professionals experienced in the specific field as well as a comprehensive comparison of individual developments.

5.4 Recommendations for Further Research

The results of this research could be used to conduct further research that analyses individual cases in depth since specific differences have been determined between Germany and the *UK*. Of particular interest could be the historical industrial buildings in more decentralised locations, where the 'creative placemaking' approach can be further analysed with the inclusion of pedestrian frequency analysis in the research. The limitation of research to mills could additionally be of interest in this context.

6 Summary and Conclusion

The study investigated the research question of which distinctions exist between Germany and the United Kingdom in their approaches to handling historical industrial buildings and potential developments. The sub-question for the conducted research was whether strategies and possibilities for redevelopment can be determined from an international comparison between Germany and the United Kingdom.

A qualitative study was conducted to determine the response to the research question. The study was conducted using expert interviews carried out online via video conference. Thereby, the interviews were conducted as semi-structured interviews. Based on the conducted surveys, it can be concluded that there are fundamental similarities between the two countries investigated, as well as distinct differences that can be identified. These differences have led to the development of a potential utilisation concept.

The differences identified relate to the origins behind the buildings, the system of heritage protection, cooperation with the heritage authorities, special financial claims, and a specific utilisation concept for the historic industrial buildings.

Commencing with the origin and the history of industrial buildings, the start of constructing the first factories varies around ninety years, while the United Kingdom started in 1760, and Germany around 1850 with the *Industrial Revolution* (Allen, 2009, p. 135; Ziegler, et al., 2012). Accordingly, the oldest buildings still to be found in England, for instance, the Shrewsbury Flaxmill, which also symbolises the beginning of iron-framed construction (Clarkson, 1990, p. 50), are significantly older than the first buildings constructed in Germany. Furthermore, the major sectors of industry and their corresponding buildings have varied throughout the centuries, up until the present time. The *UK* was the world leader

R. Groer et al., *Historical Industrial Buildings and their Real Estate Utilisation*, Studien zum nachhaltigen Bauen und Wirtschaften,
https://doi.org/10.1007/978-3-031-89417-6_6

in the wool and textile industry until the middle of the nineteenth century and therefore has a great quantity of different mill buildings, which are regularly situated along the canals due to the driving force and transport routes. The coal, iron, and steel industries were among the other existing branches (Allen, 2009, pp. 217–218). Germany, on the other hand, also began with the textile, coal, and steel industries, but subsequently developed increasingly into the mechanical engineering, electrical, and chemical industries (Ziegler, et. al, 2012, p. 51; Hahn, 2011, pp. 42–43). Due to the extensive losses in the world wars, particularly in Germany, the industrial buildings and industrial sites that remain today differ to some extent and are mostly unique and customised buildings and structures in both nations.

In terms of heritage protection, fundamental differences exist in the organisation of state structures between the *UK* with four individual nations, and Germany with sixteen federal states. In the *UK*, each of the four countries is responsible for enforcing heritage protection while in Germany, each federal state is responsible. In the *UK*, built heritage is categorised in all four nations according to categories or grades. In England and Wales, for instance, these are *Grade I*, *Grade II** and *Grade II*. The three levels descending to *Grade I* are each assigned a greater protection status, resulting in fewer opportunities for authorised changes (RUK2, RUK3, RUK4, RUK5). In addition, the *UK* has a second list *HAR* that protects particularly endangered buildings. In Germany, on the other hand, there is only one level of heritage protection, which has its own legal regulations in each federal state and is more detailed, for instance, including the protection of interiors and furnishings (RG1). What both countries have in common is the comprehensive protection of areas with an external appearance, individual buildings or parts of buildings, and the maintenance of a list of protected buildings. An authority that is a government body is established in each country within the UK. In contrast, in Germany, each federal state is responsible for the listing and generates the built heritage list independently.

An interesting approach for Germany could be the consideration of the categorisation of built heritage. Along with this is a gradation of the protected status's scope and a more or less restrictive treatment of applications for alterations. Furthermore, the approach of a heritage authority for the country with a standardised system could be considered.

Regarding the interaction with the heritage authorities, which is to be evaluated here under the remark that both the interviewees and the local authorities with their local heritage officers are each affected by subjective circumstances, it can be deduced from the interviews that it is often regarded as being very difficult.

However, the statements in this research rate the German authorities considerably lower than the *UK* heritage authorities. If the statements of the German and *UK* experts surveyed are assumed to be representative, a possible approach for better cooperation in Germany would be the *UK* method of pre-application, which cannot entirely be compared with the German 'Bauvoranfrage' since RG3 stated that planning must be almost completed but involves the authorities more intensively in the conception from the beginning and the authorities often undertake comparisons between developments and incorporate solutions from their own experience.

Regarding financial claims, there is the distinction that the *UK* has a *National Lottery Heritage Fund*, which—nomen est omen—provides money from the national lottery for instance for the preservation of built heritage. There is furthermore an exemption from property tax for unused or untenanted heritage buildings in England, for example, which could be a considerable approaches for Germany. For the *UK*, considerable adoptable approaches for financial claims that are granted in Germany could be the interest subsidy loans for heritage refurbishment, especially for sustainable heritage developments, and furthermore special depreciation for invested money in refurbishment assigned to heritage buildings instead of charging *VAT* on some refurbishment projects and not applying to specific new construction.

On the technical side, it can be noted that fundamentally comparable problems were stated in this survey with building structures, contamination, vandalism, and environmental influences by plants and protected animals occurring for both countries.

Urban development and neighbourhood development were trends that all experts from both countries mentioned with high relevance in connection with the project development and revitalisation of formerly used historical industrial properties and their locations. In this context, the experts emphasised the extraordinary significance of the bundling of architectural, historical, and social values associated with these buildings. On both sides, these characteristics were associated with the common good and values that people, especially younger generations identify with. Redevelopment of historical industrial buildings hence can generate vibrant areas while generating new living spaces, workplaces, and new neighbourhoods with associated communities.

A key driver is also identified on both sides in the field of sustainability with sustainable development of existing buildings by renovation and revitalisation as well as in connection with the historical and social aspects through the creation of described urban spaces.

Merging the characteristics mentioned, a possible utilisation concept can be derived by incorporating different elements, in accordance with the required mixed-use approach. The demanded utilisation concepts for historical industrial buildings is referred to by experts from both countries to be a mixed-use tenant structure consisting of commercial, office, and residential components. The necessity of integrating leisure experience areas with retail was emphasised by a German expert, who discussed the structural shift in the retail industry and classified this as a trend. An expert from England provided the approach of 'creative placemaking' which encompasses elements of heritage, culture, and commerce. Consequently, it is a method that has a strong relation to community and values transferred by the redeveloped historical industrial buildings. This approach is supported by other experts from the *UK*, who mention that such places convey values and an attitude that in turn attracts companies and their employees (RUK3).

If merging the 'creative placemaking approach' with the *bohemian* trend with possible tenants such as handcrafting, artistic, events, new work, leisure, and additionally added educational institutions, a tenant mix might merge for redevelopment of historical industrial buildings. This approach of the redevelopment of historical industrial buildings could thereby be suitable for buildings situated either in the urban realm, outskirt areas, or even economically weaker regions by merging the aesthetic appeal with new use concepts and attracting people.

The integration of educational institutions in historical industrial buildings such as universities is another utilisation concept that might also be considered a single tenant concept stated by one *UK* interviewee, which usually leads to a strong credit rating of the tenant (RUK5).

This utilisation concept for historic industrial buildings is strengthened by the special charm and values associated. One German expert who mentioned the structural change in the retail sector also mentioned the attraction for people and visitors which may result in an improvement in pedestrian frequency, which hence can lead to both an increase in turnover for tenants and resulting in higher leases. Problems resulting from poor floor plans and losses of lettable space might thus be compensated for by increased pedestrian frequencies.

In order to be able to successfully implement such a mixed-used concept for historical industrial sites, all experts mentioned that the area to be developed and the individual buildings must be initially investigated and understood in their unique characters. If, based on this, a reliable concept with consideration for the preservation and appreciation of the existing structures is linked to project redevelopment, a successful project redevelopment can be achieved with the early involvement of the heritage authorities, politicians, and the regional community.

In summary, as the majority of experts mentioned, the use and management of historical industrial buildings with a sensible concept is the best way of preservation and can incorporate various existing trends in the revitalisation of historical industrial buildings and thereby flourish cities as a crystallisation point and cultural anchor.

References

Ache, P., Georgi, S., Korinke, E., & Hebecker, J. (2016). *Strukturierung des sachlichen Teilmarktes wirtschaftlich genutzter Immobilien für die Zwecke der Marktbeobachtung und Wertermittlung* (Vol. 1. Ergebnisbericht (August 2016)). Berlin: Zentraler Immobilien Ausschuss e.V.

Act of Parliament. (1882). *Ancient Monuments Protection Aet, 1882.* (Vols. 45 & 46 VICT.] Ancient Monuments Protection Act, 1882. [Ch. 73.]). London: Eyre and Spottiswoode, Printers to the Queen´s most Excellent Majesty.

Acts of Parliament . (1990). *Planning (Listed Buildings and Conservation Areas) Act 1990* (Vol. UK Public General Acts1990 c. 9). London: Queen's most Excellent Majesty by and with the advice and consent of the Lords Spiritual and Temporal, and Commons, in this present Parliament.

Adler, G. (2018). *CHANGE AND VALUES IN THE EUROPEAN TOWN AND CITY SUCCESSFUL DEVELOPMENT OF THE BUILDING STOCK* (Vol. Version June 2018). (M. Scown, Trans.) Potsdam: National Contact Point for the Protection of Urban Architectural Heritage c/o complan Kommunalberatung GmbH.

Allen, R. (2009). *The British Industrial Revolution in Global Perspective* (Vol. 6th printing). Oxford: Cambridge books, Cambridge university press.

Bargery, R. (2005, no date no date). *building conservation.* Retrieved 07 15, 2024, from The Ethics of Facadism Pragmatism versus Idealism: The Ethics of Facadism Pragmatism versus Idealism.

Bau NVO. (2023, 07 03). *Bundesministerium der Justiz gesetzte-im-Internet.* Retrieved 06 2024, from Baunutzungsverordnung in der Fassung der Bekanntmachung vom 21. November 2017 (BGBl. I S. 3786), die zuletzt durch Artikel 2 des Gesetzes vom 3. Juli 2023 (BGBl. 2023 I Nr. 176) geändert worden ist: https://www.gesetze-im-internet.de/baunvo/BJNR004290962.html

Bauer, C., & Dolgan, C. (2020). TOWARDS A DEFINITION OF LOST PLACES. *Erdkunde Archive for scientific geography, Volume 74*(No. 2), 101-116.

Baum, S., Elsasser, P., Goetzke, R., & Kreins, P. (2021). Handlungsfelder der Landnutzung. In H. Gömann & J. Fick (Eds.), *Wechselwirkungen zwischen Landnutzung und Klimawandel* (pp. 72–141). Wiesbaden: Springer Spektrum.

R. Groer et al., *Historical Industrial Buildings and their Real Estate Utilisation*, Studien zum nachhaltigen Bauen und Wirtschaften,
https://doi.org/10.1007/978-3-031-89417-6

BauNVO. (2022). *BauGB Baugesetzbuch, BauNVO, PlanZV, ImmoWertV, Raumordnungsgesetz* (Vol. 53. Auflage). München: dtv Verlagsgesellschaft mbH & Co.KG—Verlag C.H. Beck oHG. Retrieved from https://www.gesetze-im-internet.de/baunvo/__8.html.

Bienert, S., & Wagner, K. (2018). *Bewertung von Spezialimmobilien—Risiken, Benchmarks und Methoden* (Vol. 2. Auflage). Wiesbaden: Springer Gabler.

Binder , A., & Rae, J. (2024, 05 05). *Encyclopedia Britannica.* Retrieved 06 2024, from Encyclopedia Britannica: https://www.britannica.com/technology/automotive-industry/Ford-and-the-assembly-line.

Blomvall, D. (no date, no date no date). *Hisroy Commons—The National Archives, UK.* (Based on wartime intelligence files from The National Archives, UK) Retrieved 06 2024, from History Commons—Bombing Britain: an air raid map: https://www.nationalarchives.gov.uk/help-with-your-research/research-guides/bomb-census-survey-records-1940-1945/.

Borchard, K., Bucher, W., Müller, W., Priebs, A., Scholich, D., Sinz, M., & Müller, P. (2011). *Grundriss der Raumordnung und Raumentwicklung.* Hannover: Akademie für Raumforschung und Landesplanung (ARL®).

bulwiengesa AG. (2014). *Market report No. 1 first half-year of 2014 Transparency on Germany's IndusTrIal real esTaTe markeT.* Berlin: Initiative Unternehmensimmobilien.

Cambridge Dictionary. (2024). *Cambridge Dictionary.* Retrieved 06 2024, from ambridge University Press & Assessment 2024: https://dictionary.cambridge.org/dictionary/english/industrial-property.

Carlin, D. (2022, 04 13). *UN environment programme | finance initiative.* Retrieved 06 2024, from 40% of emissions come from real estate; here's how the sector can decarbonize: https://www.unepfi.org/themes/climate-change/40-of-emissions-come-from-real-estate-heres-how-the-sector-can-decarbonize/#:~:text=Nearly%2040%25%20of%20global%20carbon,remaining%2030%25%20comes%20from%20construction.

Cartwright, M. (2023, 05 02). *World History Encyclopedia.* Retrieved 06 2024, from Wordl History Encyclopedia—British Industrial Revolution: https://www.worldhistory.org/British_Industrial_Revolution/#:~:text=Definition&text=The%20British%20Industrial%20Revolution%20(1760,in%20ever%2Dgrowing%20urban%20centres.

Clarkson, L. (1990). *Studies in Economic and Social History The Industrial Revolution: ACompendium.* Belfast: MACMILLAN EDUCATION LTD.

Council of Europe. (1985). *Convention for the Protection of the Architectural Heritage of Europe* * (Vols. European Treaty Series—No. 121). Granada: Council of Europe.

Dammert, B. (2019). Rechtsgrundlagen der Immobilienwirtschaft: Öffentliches und privates Baurecht. In B. Dammert, *Grundlagen der Immobilienwirtschaft Recht—Steuern—Marketing—Finanzierung—Bestandsmanagement—Projektentwicklung* (Vols. 10., überarbeitete und aktualisierte Auflage, pp. 105–191). Leipzig: Springer Gabler.

De Roos, D. (2008). *COMMERCIAL REAL ESTATE INVESTING.* New Jersey: John Wiley & Sons Inc.

Dembski, S., Sykes , O., Couch, C., Desjardins, X., Evers, D., Osterhage, F., . . . Zimmermann, K. (2021). Reurbanisation and suburbia in Northwest Europe: A comparative perspective on spatial trends and policy approaches. *Progress in Planning, Volume 150* , n. pag. .

Department for Communities and Local Government. (2014). *Plain English guide to the Planning System.* London: Crown copyright.

Department for Levelling Up, Housing and Communiti. (2023). *National Planning Policy Framework.* London: Department for Levelling Up, Housing and Communities © Crown copyright 2023.

Eisenmann, D. (2017). *Handreichung der Kultusministerkonferenz der Länder zum UNESCO-Welterbe* (Vols. UNESCO-WELTERBE Handreichung der Kultusministerkonferenz der Länder Beschluss der Kultusministerkonferenz vom 12. Oktober 2017). Berlin: Sekretariat der Kultusministerkonfererenz.

English Heritage, & bdrc continental. (2011). *Industrial Heritage at Risk Public Attitudes Survey.* no city: bdrc continental.

English Heritage. (no date, no date no date). *English Heritage.* Retrieved 06 29, 2024, from OUR HISTORY: https://www.english-heritage.org.uk/about-us/our-history/?_gl=1*195h4cs*_up*MQ..*_ga*MTg5NDQwODg5LjE3MTk3NDMzMTA.*_ga_QK86RM1N34*MTcxOTc0MzMwOS4xLjAuMTcxOTc0MzMwOS4wLjAuMA.

Ewen, S. (2016). *What is Urban History?* Cambridge: Polity.

Ewringmann, D., Perner, A., & Schön, D. (2000). *Das Europäische Raumentwicklungskonzept und die Raumordnung in Deutschland* (Vol. IzR (Informationen zur Raumentwicklung); Ausgabe 3/4.2000). Bonn: Bundesamt für Bauwesen und Raumordnung.

Föhl, A. (no date). *Industriedenkmalpflege im 20. Jahrhundert.* Wüstenrot Stiftung, DenkmalPraxisModerne. Ludwigsburg: Wüstenrot Stiftung—DenkmalPraxisModerne.

Fuß, S., & Karbach, U. (2019). *Grundlagen der Transkription Eine Praktische Einführung 2. Auflage* (Vol. 2. Auflage). Regensburg: Verlag Barbara Budrich Opladen und Toronto 2019.

Gläser, J., & Laudel, G. (2009). *Experteninterviews und qualitative Inhaltsanalyse als Instrumente rekonstruierender Untersuchungen* (Vol. 3., überarbeitete Auflage). Wiesbaden: VS Verlag für Sozialswissenschaften.

Glatte, T. (2024). *Location Strategies* (Vol. International Site and Facility Planning as part of Corporate Strategies). Heidelberg: Springer Fachmedien Wiesbaden GmbH, ein Teil von Springer Nature 2024.

Glatte, T. (2014). *Entwicklung betrieblicher Immobilien—Beschaffung und Verwertung von betrieblichen Immobilien im Corporate Real Estate Management.* Dresden: Springer Vieweg.

Glatte, T. (2021). *Corporate Real Estate Management* (Vol. essentials). Heidelberg: Springer Vieweg.

Graafland , A., Hauptmann, D., & Sassen, S. (2001). The Impact of the New Technologies and Globalization on Cities. *Cities in Transition, 5th edn.* , 650–658.

Gravagnuolo , A., Fusco Girard, L., Kourtit, K., & Nijkamp, P. (2021, 10). Adaptive re-use of urban cultural resources: Contours of circular city planning. *City, Culture and Society, Vol. 26*, 1-.

Grundsteuer. (no date, no date no date). *Das Grundsteuerportal für Deutschland.* Retrieved 07 010, 2024, from Grundsteuer: https://grundsteuer.de/steuererklaerung/anlage-befreiung-verguenstigung/denkmalschutz#:~:text=In%20allen%20Bundesländern%20werden%20Baudenkmäler,bis%20zu%2010%20%25%20niedriger%20ausfallen.

Hacking, N., & Flynn, A. (2018, 03). Protesting against neoliberal and illiberal governmentalities: Acomparative analysis of waste governance in the UK and China. *Political Geography, Political Geography 63*, 31–42.

Hahn, H.-W. (2011). *Enzyklopädie Deutscher Geschichte Band 49—Die Industrielle Revolution in Deutschland* (Vol. 3. durchgesehene und um einen Nachtrag erweiterte Auflage). München: Oldenbourg Wissenschaftsverlag GmbH.

Harzenetter, D., Haspel, P., Hesse, D.-I., & Karg, P. (2016). *Leitbild Denkmalpflege Zur Standortbestimmung der Denkmalpflege heute* (Vol. 2. Auflage, Printed in EU). Wiesbaden: Vereinigung der Landesdenkmalpfleger c/o Landesamt für Denkmalpflege Hessen.

Haynes, B., Nunnington , N., & Eccles, T. (2017). *CORPORATEREALESTATE ASSETMANAGEMENT STRATEGY AND IMPLEMENTATION* (Vol. Second edition published 2017 by Routledge). Abingdon: Routledge is an imprint of the Taylor & Francis Group.

Historic England. (2011, 10 04). *Support Your Industrial Heritage.* Retrieved 07 01, 2024, from Historic England: https://historicengland.org.uk/advice/caring-for-heritage/industrial-heritage/support-industrial-heritage/.

Historic England. (2014, no date no date). *Historic England.* Retrieved 06 30, 2024, from Scheduled Monuments A Guide for Owners and Occupiers: https://historicengland.org.uk/images-books/publications/scheduled-monuments-guide-for-owners-and-occupiers/guideforownersofscheduledmonuments/.

Historic England. (2017). *Industrial Buildings—Listing Selection Guide* (Vol. First published by English Heritage April 2011. This edition published by Historic England December 2017. All images © Historic England unless otherwise stated.). Swindon: Historic England.

Historic England. (2023, 11 09). *Historic England*, Historic England Reveals its Heritage at Risk Register 2023. Retrieved 07 01, 2024, from Historic England Reveals its Heritage at Risk Register 2023: https://historicengland.org.uk/whats-new/news/heritage-at-risk-2023/.

Historic England. (2023, 11 29). *Techniques for Valuing Heritage in Economic Terms.* Retrieved from Historic England: https://historicengland.org.uk/research/heritage-counts/heritage-and-economy/heritage-valuation/.

Historic England. (no date, no date no date). *About The List.* Retrieved 07 01, 2024, from Historic England: https://historicengland.org.uk/listing/the-list/about-the-list/.

Historic England. (no date, no date no date). *Designating and Managing a Conservation Area.* Retrieved 06 30, 2024, from Historic England: https://historicengland.org.uk/advice/planning/conservation-areas/.

Historic England. (no date, no date no date). *Historic England*, A history of listing since 1560. Retrieved 06 29, 2024, from Timeline of Conservation Catalysts and Legislation: https://historicengland.org.uk/whats-new/features/conservation-listing-timeline/.

Historic England. (no date, no date no date). *Historic England.* Retrieved 06 30, 2024, from Scheduled Monuments: https://historicengland.org.uk/listing/what-is-designation/scheduled-monuments/.

Historic England. (no date, no date no date). *Historic England.* Retrieved 06 30, 2024, from Listing Selection Guides: https://historicengland.org.uk/listing/selection-criteria/listing-selection/.

Historic England. (no date, no date no date). *Listed Buildings.* Retrieved 06 30, 2024, from Historic England: https://historicengland.org.uk/listing/what-is-designation/listed-buildings/.

Historic England. (no date, no date no date). *What is the Heritage at Risk Programme?* Retrieved 07 01, 2024, from Historic England: https://historicengland.org.uk/advice/heritage-at-risk/types/.

HM Revenue & Customs—GOV.UK. (2022, 05 09). *Guidance on capital taxation and the national heritage*, Updated 9 May 2022. Retrieved 07 01, 2024, from Government.UK: https://www.gov.uk/government/publications/capital-taxation-and-tax-exempt-heritage-assets/guidance-on-capital-taxation-and-the-national-heritage#chapter-3-capital-taxes--summary-of-heritage-reliefs-available.

Hudson, P. (1989). *Regions and Industries A perspective on the industrial revolution in Britain* (Vol. Transferred to digital printing 2004). Liverpool: Press Syndicate of the University of Cambridg.

Jackson, R. (1985). Growth and Deceleration in English Agriculture, 1660–1790. *THE ECONOMIC HISTORY REVIEW, SECOND SERIES, VOLUME XXXVIII, No. 3, AUGUST i985*, 333–351.

Josephson, M., & Conot, R. (2024, 05 27). *Encyclopedia Britannic*. Retrieved from Britannica—Thomas Edison American inventor: https://www.britannica.com/biography/Thomas-Edison..

Kiesow, G., Neuffer, M., Eberl, W., & Kirschbaum, J. (1983). *Zur Lage des Denkmalschutzes und der Denkmalpflege in der Bundesrepublik Deutschland Memorandum* (Vol. Schriftenreihe Band 20). Bonn: Deutsches Nationalkomitee für Denkmalschutz, Geschäftsstelle beim Bundesminister des Innern,chriftenreihe des Deutschen Nationalkomitees für Denkmalschutz.

King, A. D. (2015). *Global Cities—Post-Imperialism and the Internationalization of London* (Vols. Volume 5—ROUTLEDGE LIBRARY EDITIONS: ECONOMIC GEOGRAPHY). London: Routledge Taylor and Francis Group London and New York.

Kirkwood, N. (2005). *Manufactured Sites Rethinking the Post-Industrial Landscape*. London & New York: Spon Press.

Köster, G. (2021). *Projektentwicklung von Immobilien—Grundlagenwissen und Handlungsempfehlungen*—(Vol. essentials). Hamburg: Springer Gabler.

Krüper, J. (2016). Bauplanungsrecht. In N. Rottke, A. Goepfert, & K. Hamberger (Eds.), *Immobilienwirtschaftslehre Recht* (pp. 355–400). Wiesbaden: Springer Fachmedien Wiesbaden.

Leopoldsberger, G., Everling, O., & Salostowitz, P. (2022). Bewertung von Industrieimmobilien. In O. Everling, & P. Salostowitz, *Rating von Industrie- immobilien* (Vol. Edition Frankfurt School, pp. 143–152). Frankfurt: Springer Fachmedien Wiesbaden GmbH, ein Teil von Springer Nature 2023.

Liersch, A., & Asef, D. (2018). *Bildung und Kultur Spartenbericht Baukultur, Denkmalschutz und Denkmalpflege*. online: Statistisches Bundesamt (Destatis).

LogIVEST. (2024, no date no date). *LogInVest*. Retrieved from LogiVest: https://www.logivest.de/industrie/industriegebiete

Lüken, S. (2014, 09 1). *LEMO—lebendiges Museum Online*. (Deutsches Historisches Museum, Berlin) Retrieved 07 01, 2024, from Erster Weltkrieg—Der Luftkrieg: https://www.dhm.de/lemo/kapitel/erster-weltkrieg/kriegsverlauf/luftkrieg.html#:~:text=Bis%20zum%20Frühjahr%201917%20bombardierten,gab%20mehr%20als%20500%20Tote.

Meister, D., & Schulz, C. (2018). Bewertung von Industrieimmobilien. In S. Bienert, & K. Wagner, *Bewertung von Spezialimmobilien Risiken, Benchmarks und Methoden 2. Auflage*

(Vol. 2. Auflage, pp. 747–769). Regensburg: Springer Fachmedien Wiesbaden GmbH 2005, 2018.

Morgan , K. (2011). *The Birth of Industrial Britain, 1750–1850* (Vol. Second edition). Great Britain: Routledge Taylor and Francis Group London and New York.

Morri, G., & Benedetto, P. (2019). *Commercial Property Valuation.* Chichester: John Wiley & Sons Ltd.

Morrice, D. (2019). *Conservation Area Appraisal, Designation and Management—Historic England Advice Note 1 (Second Edition)* (Vol. 2nd edition updates). London: Historic England.

Morrice, D. (2021). *Historic England 2021 Listed Building Consent: Historic England Advice Note 16. Swindon. Historic England.* (Vol. Historic England Advice Note 16). London: Historic England.

Oeverman, H., & Mieg, H. (2015). *Industrial Heritage Sites in Transtormation* (Vol. ROUTLEDGE STUDIES I N HERITAGE ClashofDiscourses). New York & Abington: Routledge Taylor & Francis Group.

Pfnür, A. (2011). *Modernes Immobilienmanagement (Vols. Dritte, vollständig überarbeitete und* (aktualisierte). Darmstadt: Springer, Heidelberg Dordrecht London New York.

Pfnür, A., Eberhardt, M., & Herr, T. (2022). *Transformation der Immobilienwirtschaft—Geschäftsmodelle, Strukturen, Prozesse und Produkte im Wandel.* Wiesbaden: Springer Gabler.

Pickard, R. (2002). A comparative review of policy for the protection of the architectural heritage of Europe. *Heritage Studies, International Journal of Heritage Studies,* (Vol. 8. No. 4, Routledge pp. 349–364).

Ravin, D. (2017, 09 02). Urban-digital. Retrieved 06 26, 2024, from Stadtentwicklung erklärt: https://urban-digital.de/stadtentwicklung/.

Ritchie, H., Samborska, V., & Roser, M. (2024, 02). *Our world in Data.* Retrieved 06 2024, from Our world in Data—Urbanization: https://ourworldindata.org/urbanization.

Rukschcio, B., Nitzschke, N., Bratz, J., Kreft, L., Knufinke, J.-I., Schwope, L., & Haberle, H. (2019). *Baukultur Bericht Erbe—Bestand—Zukunft 2018/2019* (Vol. 2. Auflage). Potsdam: Bundesstiftung Baukultur BSBK Reiner Nagel.

Schmitz-Morkramer, C. (2022). All that is solid melts into air—Architektur und die Transformationen in der Immobilienwirtschaft. In A. Pfnür, M. Eberhardt, & T. Herr (Eds.), *Transformation der Immobilienwirtschaft Geschäftsmodelle, Strukturen, Prozesse und Produkte im Wandel* (pp. 461–471). Wiesbaden: Springer Gabler.

Siedentop, S. (2008). Die Rückkehr der Städte? Zur Plausibilisierung der Reurbanisierungshypothese. *Informationen zur Raumentwicklung, Heft, 3*(4), 193–210.

Society for the Protection of Ancient Buildings, S. (no date, no date no date). *The Society for the Protection of Ancient Buildings (SPAB).* Retrieved 06 29, 2024, from Different types of heritage assets: https://www.spab.org.uk/advice/different-types-heritage-assets.

Society of Property Researcheers Germany . (2007). *Wertermittlung von Baudenkmalen* (Vol. Wertermittlung von Baudenkmalen August 2007). Wiesbaden: Gesellschaft für Immobilienwirtschaftliche Forschung e.V.—Society of Property Researcheers Germany.

SPAB. (no date, no date no date). *Different types of heritage assets,* The Society for the Protection of Ancient Buildings. Retrieved 07 01, 2024, from The Society for the Protection of Ancient Buildings: https://www.spab.org.uk/advice/different-types-heritage-assets.

Statista. (no date, no date no date). *Statista.* Retrieved from Statista—Industry insights Industrial Real Estate: https://www.statista.com/markets/460/topic/2469/industrial-real-estate/#overview.

Stearns , P. (2021). *The Industrial Revolution in World History* (Vol. Fift Edition). New York: Routledge Taylor & Francis Group New York and Lonon.

Sutcliffe, A. (1988, 07). The Town Planning Review. *Britain's First Town Planning Act: A Review of the 1909 Achievement, Vol. 59* (No. 3), pp. 289–303.

Taylor and Francis Group, & The National Archives. (2019, 10 16). *Taylor & Francis an infroma business.* (Image: Crown Copyright courtesy of The National Archives. 'HO 192/626 Rocket bomb damage to Chrysler Works, London, 12 September 1944') Retrieved 06 2024, from Plotting air raids on Britain: map shows devastating impact of WWII Luftwaffe strikes: https://newsroom.taylorandfrancisgroup.com/plotting-air-raids-on-britain-map-shows-devastating-impact-of-wwii-luftwaffe-strikes/

Taylor, J., & Sorensen, G. (2021). *The Building Conservation Directory 2021* (Vols. The Twenty-eight Edition of the Directory). Tisbury: Cathdral Communication Limited. Retrieved from Heritage Protection: https://www.buildingconservation.com/articles/heritage-protection21/heritage-protection21.html

Taylor, J., & Sorensen, G. (2023). *Building Conservation Directory 2023 30th Edition* (Vol. 30th Edition). Wiltshire: Cathedral Communications Limited.

Taylor, J., & Sorenson, G. (2014, no date no date). *The Building Conservation Directory 2014.* Tisbury, The Twenty-first Edition of the Dictionary: Cathedral Communications Limited. Retrieved 05 20, 2024, from Heritage Protection in the UK: https://www.buildingconservation.com/articles/uk-heritage-protection/uk-heritage-protection.htm.

The National Lottery Heritage Fund. (no date, no date no date). *ur vision is for heritage to be valued, cared for and sustained for everyone, now and in the future.* Retrieved 07 01, 2024, from Heritage Fund: https://www.heritagefund.org.uk/about#:~:text=As%20The%20National%20Lottery%20Heritage,devolved%20governments%20to%20heritage%20organisations.

The Reigate & Banstead Borough Council. (no date, no date no date). Retrieved 06 26, 2024, from Contaminated land overview: https://www.reigate-banstead.gov.uk/info/20335/contaminated_land/434/contaminated_land_overview#:~:text=Contaminated%20land%20is%20land%20polluted,contaminated%20sites%20within%20the%20UK.

Towell, R., & Pugh, A. (2007). *Heritage Protection for the 21st Century* (By the Secretary of State for Culture, Media and Sport and the Minister for Culture, Welsh Language and Sport, Welsh Assembly Government ed., Vols. Presented to Parliament by the Secretary of State for Culture, Media and Sport By Command of Her Majesty, March 2007). London: The Stationery Office.

Turok, I., & Mykhnenko, V. (2007). *Cities—The trajectories of European cities, 1960–2005* (Vol. Volume 24). Glasgow: Department of Urban Studies and Centre for Public Policy for Regions, University of Glasgow.

UK Parliament. (no date, no date no date). *UK Parliament.* Retrieved 06 29, 2024, from Preserving historic sites and buildings: https://www.parliament.uk/about/living-heritage/transformingsociety/towncountry/landscape/overview/historicsites/#:~:text=The%20first%20legislation%20on%20the,single%20inspector%20of%20ancient%20monuments.

UK Public General Acts. (1990). *Planning (Listed Buildings and Conservation Areas) Act 1990* (Vol. up to date with all changes known to be in force on or before 18 May 2024.). London: UK Public General Acts.

UN-Habitat. (2022). *Envisaging the Future of Cities—World Cities Report 2022* . Nairobi: United Nations Human Settlements Programme (UN-Habitat).

UNEP Emissions Gap Report 2023. (2023). *United Nations*, UNEP Emissions Gap Report 2023. Retrieved 06 2024, from United Nations Climate Action: https://www.un.org/en/climatechange/net-zero-coalition.

UNESCO Constitution. (2023, 11 22). *UNESCO*. Retrieved 06 2024, from UNESCO in brief: https://www.unesco.org/en/brief.

UNESCO World Heritage Centre 1992—2024. (no date, no date no date). *Unesco World Heritage Convention*. Retrieved 06 24, 2024, from World Heritage Convention > States Parties > Germany: https://whc.unesco.org/en/statesparties/de

UNESCO World Heritage Centre 1992–2024. (no date, no date no date). *UNESCO World Heritage Convention*. Retrieved 06 24, 2024, from World Heritage Convention > States Parties > United Kingdom of Great Britain and Northern Ireland: https://whc.unesco.org/en/statesparties/gb/.

UNESCO World Heritage Centre. (2019). *Operational Guidelines for the Implementation of the World Heritage Convention UNITED NATIONS EDUCATIONAL, SCIENTIFIC AND CULTURAL ORGANIZATION INTERGOVERNMENTAL COMMITTEE FOR THE PROTECTION OF THE WORLD CULTURAL AND NATURAL HERITAGE* (Vol. WHC.19/01 10 July 2019). Paris: UNESCO World Heritage Centre.

UNESCO-World-Heritage-Centre. (no date, no date no date). *UNESCO World Heritage Concention*. Retrieved 06 2024, from World Heritage List: https://whc.unesco.org/en/list/.

Urmersbach, B. (2024). *Demographics of the UK*. Berlin: Statista. Retrieved from Statistiken zu Großbritannien: https://de.statista.com/themen/122/grossbritannien/#topicOverview.

Voigt, P. (2018, 09 27). *Wirtschaftslexikon Gabler*. Retrieved 06 2024, from Springer Gabler: https://wirtschaftslexikon.gabler.de/definition/industrielle-revolution-38116/version-334867.

Wild , M. (2024). *Denkmalschutz—Grundwissen für Denkmaleigentümer—essentials*. Münster: Springer Vieweg—Springer Fachmedien Wiesbaden GmbH, ein Teil von Springer Nature 2024.

Wild, M. (2023). *Denkmalschutz—Kompendium—Handreichung für die behördliche Praxis*. Münster: Springer Vieweg.

Winter, G., Smith, L., Cave, S., & Rehfisch, A. (2016). *Comparison of the planning systems in the four UK countries*. London: The House of Commons Library.

Wirth, A., & Schneeweiß, A. (2023). *Öffentliches Baurecht praxisnah* (Vol. 4. Auflage). Mainz: Springer Vieweg.

Wise, F., Moncaster, A., & Jones, D. (2021). Rethinking retrofit of residential heritage buildings. In *Buildings and Cities* (SPECIAL COLLECTION: RETROFITTING AT SCALE: ACCELERATING CAPABILITIES FOR DOMESTIC BUILDING STOCKS ed., Vol. Volumme 2 Number 1, pp. 495–517). ubiquity press.

Wyatt, P. (2013). *Property Valuation* (Vols. Second Edition). Chichester: John Wiley & Sons, Ltd.

Ziegler, D., Brodersen, K., Kintzinger, M., Puschner , U., & Reihardt , V. (2012). *Geschichte Kompakt—Die Industrielle Revolution* (Vol. 3. Auflage). Darmstadt: Wissenschaftliche Buchgesellschaft WBG.

Zukunftsinstitut. (2023, 12 19). *Zukunftsinstitut/Future Institute*. Retrieved 06 2024, from The Megatrends: https://www.zukunftsinstitut.de/zukunftsthemen/dossier/megatrends-en.

ZUKUNFTSINSTITUT. (2023, 12 19). *Zukunftsinstitut*. Retrieved 06 26, 2024, from The Megatrends: https://www.zukunftsinstitut.de/zukunftsthemen/dossier/megatrends-en.

Zwegers, B. (2022). *Cultural Heritage in Transition A Multi-Level Perspective on World Heritage in Germany and the United Kingdom, 1970–2020* (Vols. StudiesinArt,Heritage,LawandtheMarket 4). Maastricht: Springer Nature Switzerland AG 2022.

The manufacturer's authorised representative in the EU is Springer Nature Customer Service Centre GmbH, Europaplatz 3, 69115 Heidelberg, Germany. If you have any concerns regarding our products, please contact ProductSafety@springernature.com

Printed and bound by CPI Group (UK) Ltd, Croydon, CR0 4YY

30/09/2026

02225605-0001